Collins

KS3
Science

Revision Guide

Byron Dawson, Eliot Attridge, Heidi Foxford, Emma Poole and Ed Walsh

About this Revision & Practice book

When it comes to getting the best results, practice really does make perfect!

Experts have proved that repeatedly testing yourself on a topic is far more effective than re-reading information over and over again. And, to be as effective as possible, you should space out the practice test sessions over time.

This Revision & Practice book is specially designed to support this approach to revision and includes seven different opportunities to test yourself on each topic, spaced out over time.

Revise

These pages provide a recap of everything you need to know for each topic.

You should read through all the information before taking the Quick Test at the end. This will test whether you can recall the key facts.

> **Quick Test**
>
> 1. Name three features that both plant and animal cells have.
> 2. Which part of a plant cell helps the plant to keep its shape?
> 3. Which part of the cell controls the materials that pass in and out?

Practise

These topic-based questions appear shortly after the revision pages for each topic and will test whether you have understood the topic. If you get any of the questions wrong, make sure you read the correct answer carefully.

Review

These topic-based questions appear later in the book, allowing you to revisit the topic and test how well you have remembered the information. If you get any of the questions wrong, make sure you read the correct answer carefully.

Mix it Up

These pages feature a mix of questions for all the different topics, just like you would get in a test. They will make sure you can recall the relevant information to answer a question without being told which topic it relates to.

Test Yourself on the Go

Visit our website at **collins.co.uk/collinsks3revision** and print off a set of flashcards. These pocket-sized cards feature questions and answers so that you can test yourself on all the key facts anytime and anywhere. You will also find lots more information about the advantages of spaced practice and how to plan for it.

Workbook

This section features even more topic-based and mixed test-style questions, providing two further practice opportunities for each topic to guarantee the best results.

ebook

To access the ebook visit **collins.co.uk/ebooks** and follow the step-by-step instructions.

QR Codes

Found throughout the book, the QR codes can be scanned on your smartphone for extra practice and explanations.

A QR code in the Revise section links to a Quick Recall Quiz on that topic. A QR code in the Workbook section links to a video working through the solution to one of the questions on that topic.

Contents

Recap of Key Stage 2 Science Concepts

1 All living organisms have certain things in common.

a) Copy the table below and put a tick (✔) in the box next to the characteristics found in all living things.

Hardness	
Nutrition	
Transparent	
Flying	
Flexible	
Growth	
Reproduction	
Melting	

[3]

b) Copy and complete the table below to give definitions for the following characteristics of living things. The first one has been done for you.

Characteristic	Definition: what the word means
Respiration	Converting energy from carbohydrates and fats into energy.
Excretion	
Sensitivity	
Nutrition	

[3]

2 Materials have many different properties. Jack found these materials in his father's shed.

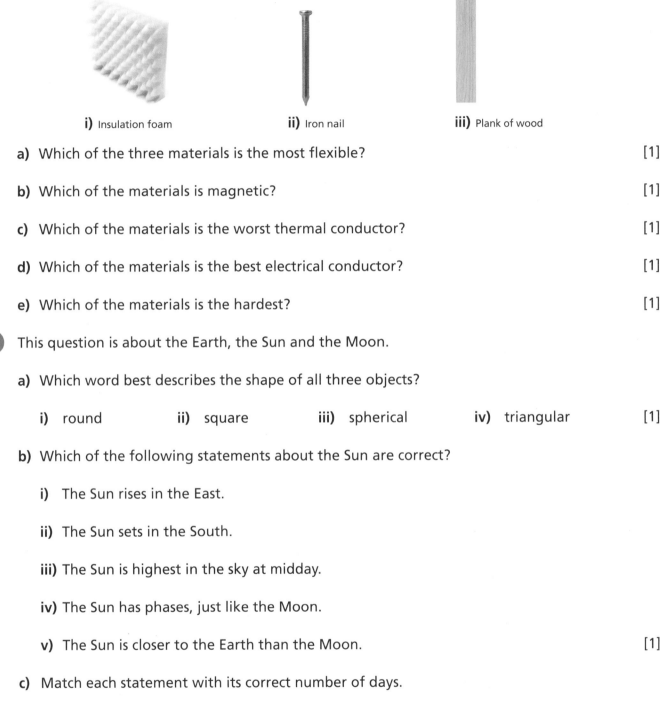

i) Insulation foam   **ii)** Iron nail   **iii)** Plank of wood

a) Which of the three materials is the most flexible? [1]

b) Which of the materials is magnetic? [1]

c) Which of the materials is the worst thermal conductor? [1]

d) Which of the materials is the best electrical conductor? [1]

e) Which of the materials is the hardest? [1]

3 This question is about the Earth, the Sun and the Moon.

a) Which word best describes the shape of all three objects?

    i) round   **ii)** square   **iii)** spherical   **iv)** triangular [1]

b) Which of the following statements about the Sun are correct?

    i) The Sun rises in the East.

    ii) The Sun sets in the South.

    iii) The Sun is highest in the sky at midday.

    iv) The Sun has phases, just like the Moon.

    v) The Sun is closer to the Earth than the Moon. [1]

c) Match each statement with its correct number of days.

Statement		Number of days
The Earth orbits the Sun		28
The Earth rotates once		1
The Moon orbits the Earth		365

[3]

Cells – the Building Blocks of Life

You must be able to:

- Use a microscope to help understand the functions of the cell
- Remember the differences between animal and plant cells
- Understand how substances move into and out of cells by diffusion
- Understand the organisation of cells.

Using a Light Microscope

- Cells are too small to see with the naked eye. Using a light microscope helps us to see and draw cells.

A plant cell drawn after observation with a light microscope

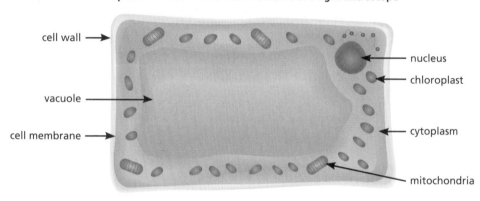

cell wall →
nucleus ←
chloroplast ←
vacuole →
cytoplasm ←
cell membrane →
mitochondria ←

Key Point
Objects need to be placed on a slide, stained and covered with a coverslip, placed on the 'stage' of the microscope, illuminated and then focussed.

How Plant and Animal Cells Work

- Animal and plant cells share some features but not others.
- Different parts of animal and plant cells have different functions.

Part	Function	Animal Cells?	Plant Cells?
Membrane	Controls what enters and leaves the cell	Yes	Yes
Cytoplasm	Place where lots of chemical reactions take place	Yes	Yes
Nucleus	Stores information (in DNA) and controls what happens in the cell	Yes	Yes
Mitochondria	Release energy from food (glucose) by aerobic respiration	Yes	Yes
Cell wall	Made from cellulose and gives rigid support to the cell	No	Yes
Vacuole	Contains a watery liquid called cell sap. It keeps the cell firm	Sometimes, but only small, temporary ones	Yes
Chloroplast	Contains green chlorophyll that absorbs light energy to allow plants to make their own food	No	Yes

Diffusion

- **Diffusion** is one of the ways that substances enter and leave cells.
- In an animal cell, oxygen and glucose diffuse through the membrane into the cell. This is because there is more oxygen and glucose outside the cell than there is inside.
- Carbon dioxide and waste products diffuse out of the cell into the blood.
- In a plant cell, carbon dioxide diffuses in. Oxygen and glucose diffuse out.

Unicellular Organisms

- **Unicellular** organisms have just one cell.
- *Euglena* has a long whip-like structure to help it move through water.
- *Amoeba* can make finger-like projections to catch food.

Organisation of Cells

- Cells of the same type carrying out the same function are usually grouped together to form a **tissue**, e.g. skin cells.
- Different types of tissue are grouped together to form **organs**, e.g. the brain.
- Different types of organs are grouped together to form **organ systems**, e.g. the nervous system.
- Different types of organ systems work together to form the organism, e.g. a human being.
- Examples of cell and organ systems include:
    - Bone cells in the skeletal system
    - Blood cells in the circulatory system
    - Nerve cells in the nervous system
    - Sperm cells in the reproductive system.

> **Key Point**
>
> Diffusing substances always move from where there is a lot of the substance (high concentration) to where there is very little (low concentration).

Euglena

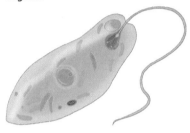

Amoeba as seen through a microscope

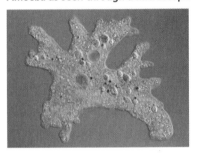

> **Key Point**
>
> cells ➡ tissues ➡ organs ➡ systems ➡ organisms

> **Key Words**
>
> membrane
> cytoplasm
> nucleus
> mitochondria
> cell wall
> vacuole
> chloroplast
> diffusion
> unicellular
> tissue
> organ
> organ system

> **Quick Test**
>
> 1. Name one structure that is found in plant cells but not animal cells.
> 2. Where in a cell is energy released from food?
> 3. Name the process where molecules move from where there are lots of them to where there are only a few.
> 4. Put these words in order of complexity starting with 'cell': cell, organism, organ, system, tissue.

Cells – the Building Blocks of Life

Quick Recall Quiz

You must be able to:

- Understand and explain the structure of the human reproductive system and how it works
- Know how reproduction and fruit dispersal works in a flowering plant
- Understand why plant reproduction is important to humans.

Reproduction in Humans

- Sexual reproduction in humans involves males and females. Males produce **sperm** cells in the **testes**. Females produce **egg cells** in the **ovary**.
- The penis deposits the sperm in the female vagina.
- Sperm swim up through the **uterus** to the oviduct.
- **Fertilisation** occurs when a sperm cell joins with an egg cell.
- The fertilised egg then grows into an **embryo** and eventually becomes a baby.

Fertilisation in animals

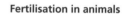
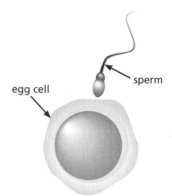

egg cell

sperm

Female reproductive system

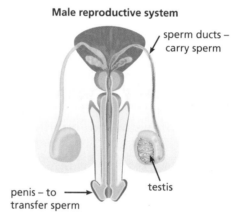

uterus

oviduct – carries egg to uterus

ovary

cervix

egg

vagina – through which baby is born

Male reproductive system

sperm ducts – carry sperm

penis – to transfer sperm

testis

Menstrual Cycle

- Females have a menstrual cycle lasting for about 28 days. This is called **menstruation**.
- On days 1–5, if pregnancy has not occurred, the uterus lining breaks down, tissue and blood are lost, and is replaced with new tissue.
- Fertilisation can only occur on or around day 14 when an egg is released from the ovary.

Gestation

- **Gestation** is the process of the embryo developing in the womb.
- The growing baby receives food and oxygen from the mother's blood through the placenta and umbilical cord.
- Therefore, if the mother smokes or drinks alcohol the baby will also receive some of the alcohol or nicotine.
- In humans, gestation ends after nine months with the birth of the baby.

> **Key Point**
>
> A human foetus takes 38 weeks to grow from the cell being fertilised to a baby.

Reproduction in Flowering Plants

- Some flowers are insect pollinated, e.g. a rose.
  - Insects visit flowers to collect sweet **nectar**
  - They transfer pollen from the **anther** of one flower to the **stigma** of another flower
  - The male pollen fertilises the female egg cell.
- Some flowers are wind pollinated, e.g. grass.
  - Wind blows pollen from one flower to another
  - Wind pollinated flowers do not have a scent or nectar and petals are not brightly coloured as they do not need to attract insects
  - They have a feathery stigma to catch the pollen.

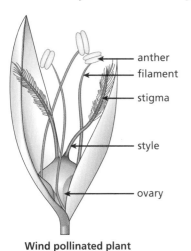

Wind pollinated plant

anther
filament
stigma
style
ovary

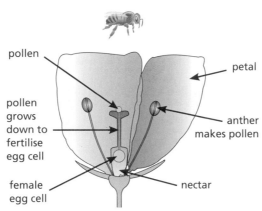

Insect pollinated plant

pollen
petal
pollen grows down to fertilise egg cell
anther makes pollen
female egg cell
nectar

Dispersal

- After fertilisation, seeds develop inside fruits. These then must be spread over a large area by **dispersal**.
- Some fruit and seeds are spread by animals, e.g. some seeds have hooks which stick to an animal's fur.
- Some are spread by wind. These often have wings or parachutes to be carried by the breeze, e.g. sycamore and dandelion seeds.
- Plants produce many seeds as most fail to grow into a new plant.

The Importance of Plant Reproduction

- Plants provide us with most of our food.
- Without insects to pollinate the flowers, many of us would starve due to lack of food.

Key Words

sperm
testes
egg cell
ovary
uterus
fertilisation
embryo
menstruation
gestation
nectar
anther
stigma
pollination
dispersal

Quick Test

1. Which two cells join together at fertilisation in humans?
2. On which day of the menstrual cycle is a female egg released?
3. Write down the differences between an insect pollinated flower and a wind pollinated flower.
4. List two ways that fruits and seeds can be dispersed.

Eating, Drinking and Breathing

Quick Recall Quiz

You must be able to:

- Know and explain how humans move air into and out of lungs
- Know and understand how oxygen and carbon dioxide move between the blood and the lungs
- Understand the effect of exercise, asthma and smoking on the breathing systems.

Breathing

- Breathing involves moving air into and out of the lungs.

When breathing in:

1. Ribs move up and out
2. **Diaphragm** flattens and moves down
3. Space inside the lungs increases
4. This increases the volume and reduces the pressure
5. Air rushes into the lungs from outside.

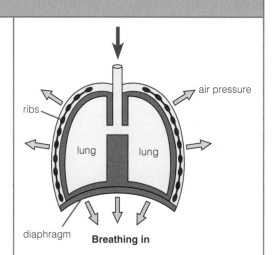

air pressure

ribs

lung   lung

diaphragm   **Breathing in**

When breathing out:

1. Ribs move down and in
2. Diaphragm moves up
3. Space inside the lungs decreases
4. This decreases the volume and increases the pressure
5. Air is pushed out of the lungs.

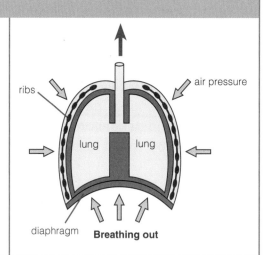

air pressure

ribs

lung   lung

diaphragm   **Breathing out**

> **Key Point**
>
> When we breathe in, air is pushed in by **air pressure** from the outside.

Gas Exchange

- The lungs are made of millions of tiny air sacs called **alveoli**.
- These air sacs are:
  - Thin
  - Moist
  - Have a good blood supply
  - Have a large surface area.

Structure of alveoli

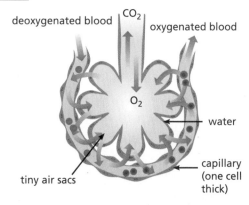

deoxygenated blood CO_2 oxygenated blood

O_2

water

tiny air sacs

capillary (one cell thick)

- Gas exchange is when:
  - Carbon dioxide leaves the blood and enters the lungs to be breathed out
  - Oxygen leaves the lungs and enters the blood.
- Gas exchange happens through the thin walls of the air sacs.
- The exchange happens because of diffusion (see page 4).

(see page 4)

The breathing system

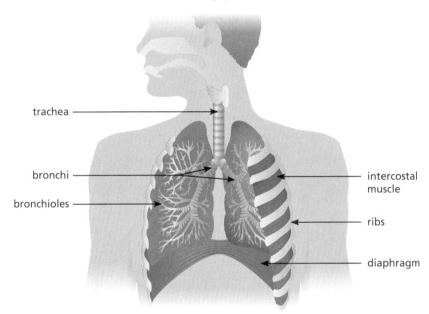

trachea

bronchi

bronchioles

intercostal muscle

ribs

diaphragm

> **Key Point**
>
> Diffusion is the movement of particles from a high to a low concentration.

Things that Affect our Breathing

Things that Affect Breathing	Effect on Breathing
Exercise	• Increases strength of the diaphragm and intercostal muscles. • Improves gas exchange. • Increases vital lung capacity (the volume of air that can be forcibly exhaled after inhaling fully).
Asthma	• Causes breathing tubes (bronchioles) to narrow, making breathing difficult.
Smoking	• Damages the breathing tubes so that mucus builds up. This causes a cough, makes breathing more difficult and makes infections more likely. • Damages the cilia that line the breathing tubes so that mucus builds up. • In the long term can cause emphysema and lung cancer.

> **Quick Test**
>
> 1. Explain how the ribs and diaphragm move to make you breathe in.
> 2. Explain what happens to the volume and pressure inside the lungs when the ribs move down and in.
> 3. Name the process by which oxygen moves from the air in our lungs into the blood.
> 4. Describe the effect smoking has on the lungs.

> **Key Words**
>
> diaphragm
> air pressure
> alveoli
> asthma

Eating, Drinking and Breathing

You must be able to:

- Explain what is meant by a healthy diet
- Explain the energy content of a healthy diet and understand what happens when a healthy diet becomes unbalanced
- Know and explain the jobs of different parts of the digestive system.

A Healthy Diet

- A healthy diet contains all the right proportions of the following substances:

Content of Healthy Diet	Purpose
Carbohydrate	Gives the body energy
Fat	Provides reserve energy supply and insulation
Protein	Used for growth
Vitamins	Needed to keep our bodies healthy.
Minerals	Needed to keep our bodies healthy.
Fibre	Helps undigested food pass quickly through the gut
Water	Dissolves chemicals so that chemical reactions can take place

- A healthy diet also contains sufficient food to provide us with just the right amount of energy.
- Energy in food is measured in calories or joules.
- 1 calorie = 4.2 joules
- A young man needs about 2500 kcal per day.
- 2500 kcal × 4.2 joules = 10,500 kJ per day.

> ### Key Point
>
> When dieticians talk about calories in food they really mean kilocalories. A kilocalorie is 1000 calories.

An Unbalanced Diet

- Eating an unbalanced diet can cause many problems:

Cause	Problem
Eating too much	Obesity
Eating too little	Starvation/malnutrition
Not eating enough protein	**Kwashiorkor**, an illness caused by severe protein deficiency. It is mostly seen in developing countries.
Not eating enough vitamins	A lack (or deficiency) of different vitamins causes different diseases, e.g. a lack of vitamin C causes **scurvy**.
Not eating enough minerals	A lack of iron causes **anaemia**. A lack of calcium causes soft bones.

The Digestive System

- The digestive system processes food that is eaten in the mouth. Food travels through the **oesophagus**, **stomach**, **intestine** and **rectum** until the waste is eliminated from the **anus**.
- Enzymes are proteins which act as what are called 'biological catalysts' that speed up reactions. During chemical digestion, enzymes break down carbohydrates, proteins and fats into smaller molecules so they can be absorbed into the blood.

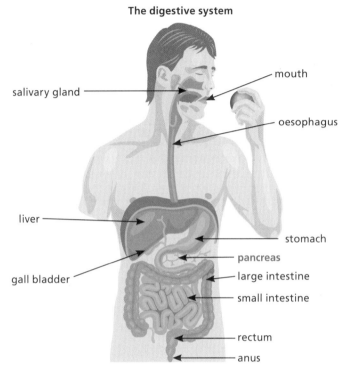

The digestive system

- salivary gland
- mouth
- oesophagus
- liver
- stomach
- pancreas
- large intestine
- gall bladder
- small intestine
- rectum
- anus

Food in Plants

- Unlike animals that eat food, plants make their own food.
- The process is called **photosynthesis**.
- Plants take water and minerals from the soil.
- They take carbon dioxide from the air.
- They use energy from the Sun to convert these substances into carbohydrates in their leaves:

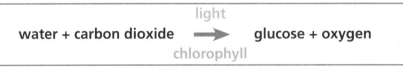

light

water + carbon dioxide ➡ **glucose + oxygen**

chlorophyll

Quick Test

1. Name five components of a healthy diet.
2. Name three possible consequences of eating an unbalanced diet.
3. Write down the different parts of the digestive system in the order food travels through them. Start with **mouth**.
4. Describe the difference in feeding between plants and animals.

Key Point

Animals eat food, plants make it.

Key Words

carbohydrate
fat
protein
vitamins
minerals
fibre
kwashiorkor
scurvy
anaemia
oesophagus
stomach
intestine
rectum
anus
enzyme (biological catalyst)
pancreas
photosynthesis

Cells – the Building Blocks of Life

1   Match the part of a cell to its function.

Part of cell		Function
Membrane		Uses light energy to produce food
Cytoplasm		Keeps the cell firm
Nucleus		Supports the cell
Mitochondria		Releases energy from glucose
Cell wall		Stores information and controls the cell
Vacuole		Where chemical reactions take place
Chloroplast		Controls what enters and leaves a cell

[7]

2   These plant cells were seen using a microscope. Make a labelled drawing of one of them.

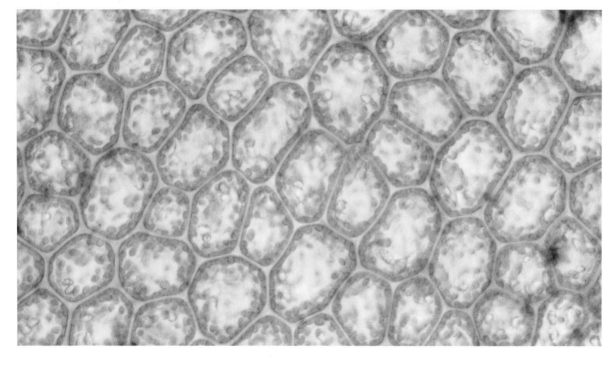

[6]

3   Which of these words describes how substances can enter or leave a cell?

   i)   cytoplasm   **ii)**   vacuole   **iii)**   diffusion   [1]

Eating, Drinking and Breathing

1. Humans need to eat a healthy diet.

   a) Explain what is meant by a healthy diet. [2]

   b) Carbohydrates, protein and fibre are all important parts of a healthy diet.

   Explain the function of each of these food types. [3]

2. Look at the diagram of the digestive system.

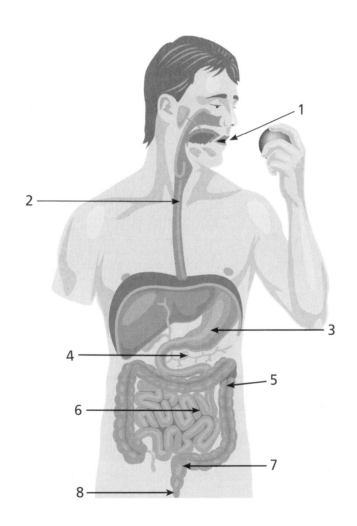

   a) Give the correct names for each part 1–8. [8]

   b) Explain what each of the parts you have labelled does. [8]

3. Explain the difference between feeding in animals and feeding in plants. [4]

Getting the Energy your Body Needs

Quick Recall Quiz

You must be able to:

- Explain what respiration is
- Understand and explain aerobic and anaerobic respiration, including the differences between them.

Respiration

- **Respiration** is the process by which organisms release energy from food.
- The energy is needed to power all the chemical processes necessary for life.
- There are two types of respiration, **aerobic** and **anaerobic**.

Aerobic Respiration

- Humans release energy from **glucose** and **oxygen** by aerobic respiration.
- Carbon dioxide and water are produced as waste products.

> **glucose + oxygen ➝ carbon dioxide + water +** energy

Anaerobic Respiration

- Anaerobic respiration takes place in humans when not enough oxygen is available.
- Humans can break down glucose into **lactic acid**.
- Less energy is released during anaerobic respiration.
- Lactic acid is also released. This quickly causes muscle pain and fatigue.
- 'Getting the burn' is when muscles produce lactic acid in anaerobic respiration.

> **glucose ➝ lactic acid +** energy

- Yeast is a microorganism that can also respire without oxygen (anaerobic respiration). Yeast breaks glucose down into alcohol and carbon dioxide.
- This process is called **fermentation**.

> **glucose ➝ alcohol + carbon dioxide +** energy

> **Key Point**
>
> Anaerobic respiration can happen without oxygen, but it can only happen for a very short time. It happens when we need a lot of oxygen very quickly, such as when we run a fast race.

> **Key Point**
>
> Fermentation is used to produce alcoholic drinks such as wine and beer.

Similarities and Differences between Aerobic and Anaerobic Respiration

	Aerobic	Anaerobic
Uses glucose	✔	✔
Uses oxygen	✔	✗
Produces carbon dioxide	✔	✔ Fermentation in yeast ✗ but not in humans
Produces water	✔	✗
Releases *lots* of energy	✔	✗
Can produce lactic acid	✗	✔ In humans ✗ but not by fermentation
Can produce alcohol	✗	✔ Fermentation in yeast ✗ but not in humans
Causes muscle fatigue	✗	✔

Quick Test

1. Name the type of respiration that uses oxygen.
2. Name the type of respiration that releases the most energy.
3. Give the type of respiration that can produce lactic acid.
4. Name the type of respiration that is needed to produce alcohol for alcoholic drink production.
5. Okan runs very fast up a short and steep hill. By the time he gets to the top his leg muscles feel very tired and begin to hurt. Explain why.

Getting the Energy your Body Needs

You must be able to:

- Explain the structure and function of the human skeleton
- Explain how muscles make the skeleton move.

The Human Skeleton

- The human skeleton has several different functions:
  - Supports the body and gives it shape.
  - Acts as a framework that enables muscles to move the body.
  - Protects parts of the body, for example, the skull protects the brain and the ribs protect the heart and lungs.
  - Makes red blood cells in the marrow of the long bones, for example, humerus and femur.

Bone marrow

Human skeleton

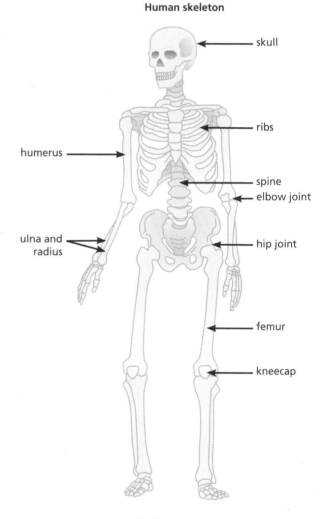

skull
ribs
humerus
spine
elbow joint
ulna and radius
hip joint
femur
kneecap

Joints and Muscles

- Bones in the skeleton are held together in **joints**.
- Joints allow the skeleton to move.
- The bones in a joint are held together by **ligaments** using the muscles that surround the joint.
- The end of each bone is covered in **cartilage** for a smooth surface that cushions the joint.
- The joint is filled with a fluid that lubricates the joint.

The knee

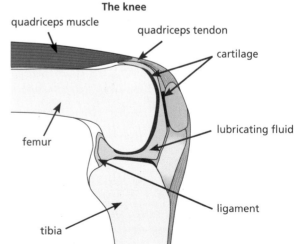

quadriceps muscle
quadriceps tendon
cartilage
femur
lubricating fluid
ligament
tibia

Muscles and Force

- Muscles move joints. Muscles are attached to bones by **tendons**.
- Each joint needs two muscles to make it work. This is called an **antagonistic pair**.
- One muscle moves the joint in one direction. The other muscle moves the joint in the opposite direction.
- Muscles work by contracting and getting shorter in length. This pulls the bone and moves the joint.

Muscles in the arm

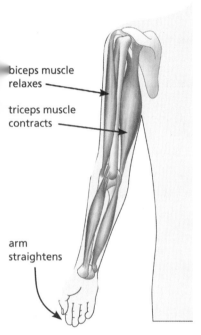
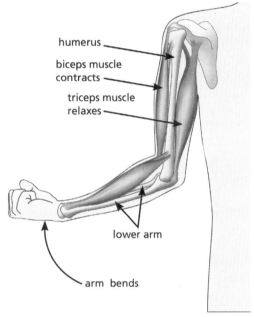

biceps muscle relaxes

triceps muscle contracts

arm straightens

humerus

biceps muscle contracts

triceps muscle relaxes

lower arm

arm bends

Revise

Key Point

Some muscles work in pairs called antagonistic pairs. When one muscle contracts the other muscle relaxes.

Key Point

Remember, a 'moment' is a scientific word used to describe the turning effect of a force.

- The force exerted by muscles can be measured in Newtons.
- To work out the turning moment, you calculate force × distance.
- Knowing the moment on the lower arm enables you to calculate the moment in the upper arm.
- So, in the diagram (right), the moment for the lower arm is:

Note that mass is converted to weight and cm converted to m.

$100\,N \times 0.3\,m = 30\,Nm$ ◄───

- With the upper arm, we know the distance the force is applied over. We also know the moment, but not the force:

Force × 0.05 m = 30 Nm

Rearranged to: Force = $\dfrac{30\,Nm}{0.05\,m}$

Therefore, the force = 600 N

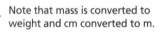

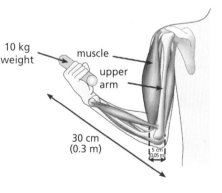

10 kg weight

muscle

upper arm

30 cm (0.3 m)

5 cm 0.05 m

Quick Test

1. Give four functions performed by the skeleton.
2. Name the tissue that attaches bones to each other in a joint.
3. Name the tissue that attaches muscle to bone.
4. Explain what antagonistic means.

Key Words

joint
ligament
cartilage
tendon
antagonistic pair

Biology

Looking at Plants and Ecosystems

You must be able to:

- Explain how photosynthesis takes place
- Understand how a green leaf is adapted for photosynthesis
- Understand the importance of photosynthesis to other living things.

Photosynthesis

- **Photosynthesis** is the process by which green plants make food.
- Green plants absorb energy from sunlight.
- They use light energy from the sun to react water with carbon dioxide to make **glucose**.
- The energy from the sunlight becomes stored in the glucose.
- **Oxygen** is released as a waste product.
- Plants use a green chemical called **chlorophyll** inside **chloroplasts** to perform photosynthesis.

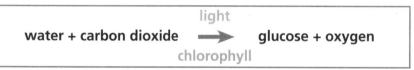

light
water + carbon dioxide ➡ **glucose + oxygen**
chlorophyll

> ### Key Point
>
> When you write down the word equation for photosynthesis, always include light and chlorophyll.

Leaves and Photosynthesis

- Leaves are the plant's factory where photosynthesis takes place.
- Leaves are adapted to do this job because they:
  - Are thin – this stops the leaves from being heavy, enabling trees to have more leaves and therefore a larger surface area
  - Have a large surface area – to catch as much sunlight as possible
  - Are green, because of the chemical chlorophyll they need in order to photosynthesise
  - Have small holes called **stomata** mostly found on the underside of the leaf. The stomata allow gases such as carbon dioxide and oxygen to enter or leave the leaf. A single hole is called a **stoma**
  - Have tiny tubes called xylem to carry water and minerals up from the roots to the leaves
  - Have tiny tubes called phloem to carry glucose away for storage.

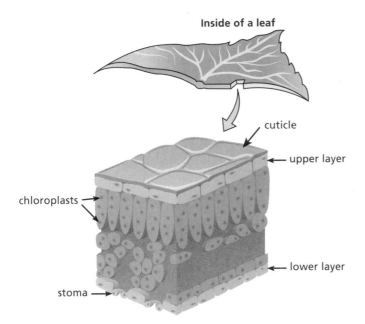

Inside of a leaf

cuticle

upper layer

chloroplasts

lower layer

stoma

The Importance of Photosynthesis

- Plants and animals depend on each other for survival – they are **interdependent**.
- For example, plants produce oxygen that animals need for respiration. Animals also need plants for food and shelter.
- Plants need animals for seed dispersal and pollination. Plants also use the carbon dioxide animals produce.
- Photosynthesis builds up complex glucose molecules from simple molecules (water and carbon dioxide). This stores energy.
- Respiration breaks down complex glucose molecules into simple molecules and releases energy.

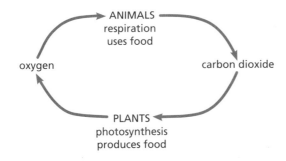

ANIMALS
respiration
uses food

oxygen

carbon dioxide

PLANTS
photosynthesis
produces food

- Most animals on Earth depend on plants for glucose, and oxygen to breathe.
- Most glucose and oxygen are produced by plants in the rain forest and algae in the oceans.

> **Key Point**
>
> Photosynthesis and respiration are the opposite of each other.
>
> - Respiration uses food and oxygen, and produces carbon dioxide.
> - Photosynthesis uses carbon dioxide, and produces food and oxygen.

> **Key Words**
>
> **photosynthesis**
> **glucose**
> **oxygen**
> **chlorophyll**
> **chloroplast**
> **stomata**
> **stoma**
> **interdependent**

Quick Test

1. Write down the word equation for photosynthesis.
2. State how photosynthesis is different to respiration.
3. Suggest a reason why stomata are found on the leaves of a plant.
4. Explain how plants and animals depend upon one another.
5. Explain what interdependence means.

Looking at Plants and Ecosystems

Quick Recall Quiz

You must be able to:

- Use food webs to explain relationships between different organisms
- Understand how organisms are affected by the environment
- Understand how differences between organisms help them to survive.

Humans and Their Food Supply

- A good food supply is important for humans.
- This food supply depends on how organisms transfer energy from one to another.
- This means that organisms in an environment are interdependent in many ways.
- Insects **pollinate** flowers so seeds and fruit can grow and be used as food by other animals.
- Humans rely on insect pollinators for many of our crops.
- Insecticides can kill harmful insects and pests, but can also kill useful pollinators.

Interdependence of Organisms

- The best way to show how organisms depend on one another is to draw a **food web**.
- Food webs are made up from many different **food chains**. They are all interdependent.
- A food chain describes what eats what in a community.
- Interdependence describes how all the living organisms in an **environment** depend upon one another.
- Food webs show how organisms depend upon one another for food; they show relationships between organisms.
- **Producers** are plants. They produce food by photosynthesis.
- **Consumers** are animals. They consume food for energy.

> **Key Point**
>
> The arrows in a food web show how energy is transferred as food from one organism to another.

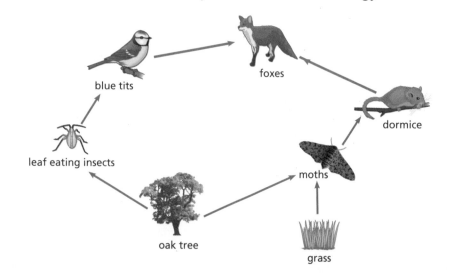

Organisms and the Environment

- Sometimes poisonous waste can get into an ecosystem:
  1 Plants at the bottom of the food web absorb the poison.
  2 The poison is passed on to the animals that feed upon them.
  3 Because these animals eat lots of plants they absorb more of the poison.
  4 The poison accumulates as it is passed up the food web; this is called **bioaccumulation**.
  5 Eventually there is enough poison in the animals at the top of the food web to kill them:

Poison in a food web

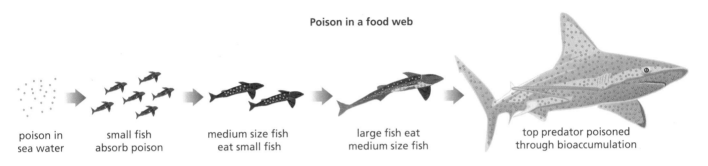

| poison in sea water | small fish absorb poison | medium size fish eat small fish | large fish eat medium size fish | top predator poisoned through bioaccumulation |

- Bioaccumulation occurs because there are many more organisms at the bottom of a food chain and only a few predators at the top.
- This forms a **pyramid of numbers** that means poisons accumulate in greater numbers in the top predators.

> **Key Point**
>
> Poison builds up as you go up the food chain. This is bioaccumulation.

Variation Between Organisms

- Most living organisms are different from one another. This is called **variation**.
- Different **species** have different characteristics.
- Different species survive in the same ecosystem because they are adapted to survive in different parts of the ecosystem.
- For example, different parts of the ecosystem are called **niches**.
- Fish live in water so they have gills; birds fly, so they have wings.
- Variation means that certain members of a species are more likely to survive when the environment changes.

> **Key Words**
>
> pollinate
> food web
> food chain
> environment
> producers
> consumers
> bioaccumulation
> pyramid of numbers
> variation
> species
> niches

> **Quick Test**
>
> 1. What do arrows on a food web show?
> 2. What does bioaccumulation mean?
> 3. Explain why variation between different organisms is important.

Cells – the Building Blocks of Life

1. Unicellular organisms have different structures from each other.

   Explain why. [2]

2. Match the type of cell with the correct organ system.

   Type of Cell

   | Bone cell |
   | Red blood cell |
   | Nerve cell |
   | Sperm cell |

   Organ system

   | Circulatory system |
   | Skeletal system |
   | Reproductive system |
   | Nervous system |

   [4]

3. Copy and complete this table by putting a ✔ and a ✗ next to each part of the reproductive system.

Part	Male	Female
Testis		
Egg cell		
Vagina		
Sperm		
Penis		

   [5]

4. Describe three differences between insect and wind pollinated flowers. [3]

5. Insects are important for our food supply.

   Explain why. [2]

6. Pollination requires the transfer of pollen from one flower to another.
   This means the grains have to be very small.

   Describe how you would use a light microscope to look at pollen grains. [6]

Eating, Drinking and Breathing

1 Look at the diagram. It shows how we breathe in and out. Use the diagram to explain what is happening when we:
- breathe in [3]
- breathe out [3]

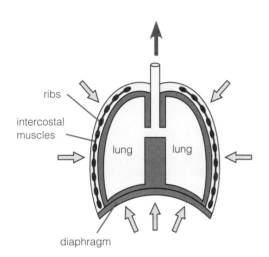 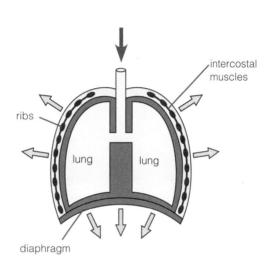

2 Regular exercise has some beneficial, long term effects on the breathing system. Describe two long term effects of regular exercise on the breathing system. [3]

3 Explain why a balanced diet should contain each of the following things:
carbohydrate, fat, protein, vitamins, minerals, fibre, water [7]

4 Explain the importance of bacteria in the digestive system. [2]

5 Chemical digestion requires digestive enzymes. Describe the function of digestive enzymes. [2]

6 The following statements describe the roles of some organs associated with the digestive system.
One of the statements is incorrect.
Copy the table and put a cross (✗) against the incorrect statement.

The oesophagus joins the mouth to the stomach	
The stomach produces an acid to break down food	
The small intestine is where water is absorbed	
Waste material leaves the body through the anus	

[1]

Getting the Energy your Body Needs

1   Copy and complete the table by putting a tick (✔) in the correct box next to each statement about respiration.

	Aerobic	Anaerobic
Uses oxygen		
Produces lactic acid		
Produces alcohol		
Releases the most energy		
Fermentation uses this type of respiration		

[5]

2   Anaerobic respiration in yeast is different to anaerobic respiration in humans. Describe the differences. [3]

3   Explain the importance of having a skeleton. [4]

4   Look at the diagram of the skeleton.

Complete the labels to give the name of the bone or joint. [10]

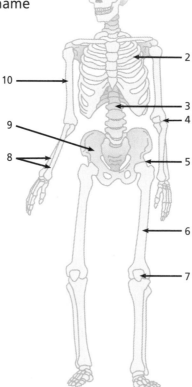

5   Explain why each joint needs at least two different muscles. [3]

Looking at Plants and Ecosystems

1 Copy and complete the table by putting a tick (✔) in the correct box next to each statement about respiration and photosynthesis.

	Respiration	Photosynthesis
Produces oxygen		
Produces carbon dioxide		
Uses energy from sunlight		
Releases energy		
Requires chlorophyll		

[5]

2 Copy and complete the table to show how leaves are adapted for photosynthesis. The first one has been done for you. [5]

Structural Adaptation	Explanation
Leaves have a large surface area.	To catch as much light as possible.
Leaf has tiny holes called stomata.	
Leaves have xylem tubes.	
Leaf cells near top of leaf contain lots of chloroplasts.	

[3]

3 a) The diagram below shows a food web. Use the food web to construct two possible food chains, starting with the oak tree.

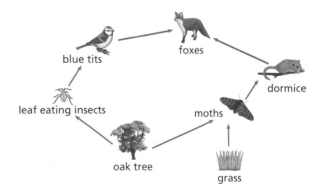

[2]

b) Name the producer in this food chain. [1]

4 Explain why insects are so important in the production of food for humans. [2]

5 Poisonous substances sometimes get released into the environment. These poisons can be more dangerous to animals at the top of the food chain. Explain why. [2]

Variation for Survival

Quick Recall Quiz

You must be able to:

- Explain how genetic information is inherited from our parents
- Understand the job of chromosomes and genes
- Know the contribution of different scientists to our understanding of DNA.

How Genetic Information is Inherited

- We inherit half our genetic information from our mother and half from our father. This is called **heredity**.
- The **inheritance** of genetic information happens when a sperm from the father fertilises an egg from the mother.

Fertilisation of egg

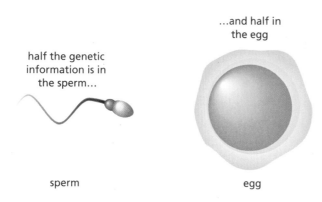

half the genetic information is in the sperm...

...and half in the egg

sperm

egg

> ### Key Point
>
> Half our genes come from our father and half from our mother. This is why we have similarities to both our mum and dad.

- The genetic information is stored on **chromosomes**, found in the nucleus of our cells.
- The nucleus of almost every cell in our body contains 46 chromosomes.
- However, sperm and eggs only contain 23 chromosomes.
- So 23 chromosomes come from our mother and 23 from our father.
- This produces new offspring with some features inherited from our mother and some from our father.

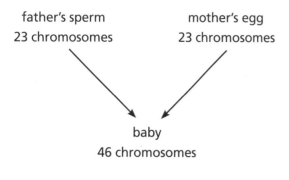

father's sperm
23 chromosomes

mother's egg
23 chromosomes

baby
46 chromosomes

- Each chromosome consists of a very long strand of **DNA**.
- The DNA is divided up into single units called **genes**.
- Each gene is a single section of DNA that codes for a protein.

From cell to gene

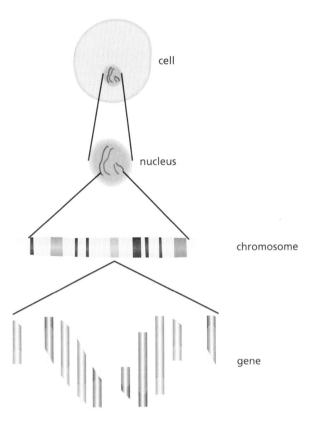

cell

nucleus

chromosome

gene

Revise

Key Point

A gene is an instruction. It tells the cell how to make a specific protein needed by the cell.

A gene

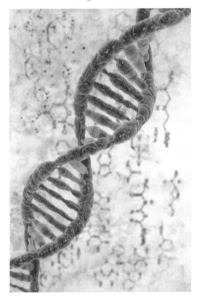

Famous Scientists and DNA

- Several different scientists played a part in discovering how genetic information is inherited.
- In 1953, James Watson and Francis Crick developed a theory for the structure of DNA.
- Maurice Wilkins helped produce evidence to support Watson and Crick's theory of the structure of DNA.
- Rosalind Franklin made X-ray images of DNA which showed that DNA was a double helix.
- Watson, Crick and Wilkins were awarded the Nobel prize for their work. Sadly, Franklin had died several years earlier and so could not be awarded the prize, even though her work was crucial to the discovery of DNA structure.

Key Point

Most scientific discoveries are a result of scientists working together and building on ideas from other scientists.

Quick Test

1. Where does a baby get its 46 chromosomes from?
2. Which is larger, a chromosome or a gene?
3. Define the term 'gene'.
4. Name four scientists who were involved in the discovery of DNA.

Key Words

heredity
inheritance
chromosome
DNA
gene

Variation for Survival

Quick Recall Quiz

You must be able to:

- Explain why variation within a species is so important
- Understand different types of variation
- Explain the effect of a changing environment on our survival.

Variation Between Species

- A **species** is a group of organisms that can reproduce to produce **fertile** offspring.
- All species are different. Scientists call this **variation**.
- Variation between species is called **interspecific variation**.

Variation Within a Species

- As well as variation between species, variation occurs *within* a species. This is called **intraspecific variation**.
- Apart from twins, all humans look different.
- Variation occurs because of sexual reproduction.
- Sexual reproduction mixes up the genes from mum and dad and this causes variation.
- Apart from identical twins, no two brothers or sisters will inherit the same combination of genes from their parents.

> **Key Point**
>
> Variation is due to different organisms having a different combination of genes.

Types of Variation

- There are two types of variation within a species: **continuous variation** and **discontinuous variation**.
- Height is an example of continuous variation. Some people are tall, others are short. But most people are somewhere in between.

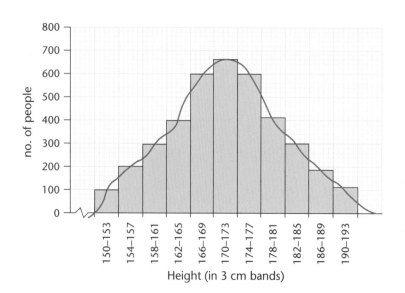

Blood groups are an example of discontinuous variation. People are either A, B, AB or O. There is no gradual range that goes from one to another.

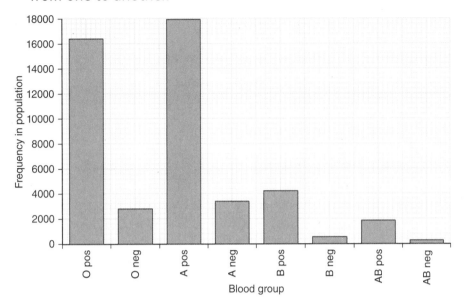

The Effect of a Changing Environment

- Variation is very important when the environment changes.
- **Extinction** of a species can happen when the environment changes. For example, global warming may cause some species to become extinct.
- This is because they are less able to compete for resources and reproduce in the changing environment.
- However, if the changes to the environment are small or occur slowly, because the members of a species are all slightly different some will be able to survive.
- These individuals survive and breed to produce new offspring, ensuring the survival of the species.
- This is why biodiversity is so important.
- The more biodiversity, the less likely it is that extinction will happen.

Gene Banks

- A gene bank is a place where scientists store seeds and cells from as many different organisms as possible.
- This helps to ensure that no genes are lost during extinction.
- These genes may be helpful in the future to provide new medicines or food.

Key Point

Biodiversity is a measure of the amount of variation between different organisms.

Key Words

species
fertile
variation
interspecific variation
intraspecific variation
continuous variation
discontinuous variation
extinction
biodiversity

Quick Test

1. Name and describe two different types of variation within a species.
2. Write down the main cause of variation in humans.
3. Define the term 'biodiversity'.
4. What is a gene bank?

Our Health and the Effects of Drugs

Quick Recall Quiz

You must be able to:

- Recall the four main types of drugs
- Describe legal and illegal recreational drugs and their effects
- Describe the dangers of smoking and drinking alcohol
- Understand drug addiction and withdrawal.

Main Types of Drugs

- A **drug** is a substance that affects the body in some way.
- There are many different drugs that affect the body in different ways.
- The four main types of drug are painkillers, depressants, stimulants and hallucinogens:

Type of drug	Effects	Examples	Dangers
Painkillers	Reduce pain and inflammation	Aspirin Paracetamol	Excess may damage stomach lining
Depressants	Make a person feel relaxed and drowsy	Cannabis	Drowsiness and lack of coordination; can cause post-use hallucinations
		Heroin	Reduces breathing and can cause death
Stimulants	Make a person feel energetic and alert	Cocaine Amphetamines	Can cause aggression and paranoia Can cause depression and panic
Hallucinogens	Make a person hear and see things more intensely	LSD	Hallucinations; users see and hear things that are not there
		Magic mushrooms	Can cause a bad trip; can cause flashbacks for some time afterwards

- All of these types of drug can be addictive.
- All drugs have **side-effects**.
- Side-effects are unwanted symptoms caused by taking the drug, e.g. rashes, headaches, or nausea.

Recreational Drugs

- Drugs that are taken for non-medical reasons are called **recreational drugs**.
- Some recreational drugs are legal and are in common use, for example, caffeine, tobacco, and alcohol.
- Some recreational drugs are illegal and can have dangerous side-effects, for example, cannabis, ecstasy and cocaine.

Smoking and Drinking Alcohol

- Two common recreational drugs are:
  - alcohol (found in alcoholic drinks such as wine and beer)
  - tobacco (found in cigarettes).
- People smoke cigarettes and drink alcohol as a way of relaxing and feeling more confident.
- Smoking and drinking too much alcohol carry serious health risks:

Smoking	Tobacco contains nicotine, a drug which speeds up heart rate and raises blood pressure, leading to increased risk of heart disease, heart attacks and strokes
	Smoking damages blood vessels and the lungs, leading to coughs, lung infections such as bronchitis, emphysema and lung cancer
Drinking alcohol	In the short term, alcohol: – slows down reactions – reduces coordination – can alter people's behaviour. In the long term, too much alcohol can cause: – liver failure – brain damage – increased risk of strokes and heart attacks – anxiety and depression

Addiction and Withdrawal

- **Addiction** to a drug means that a person feels they need to keep taking the drug and when they stop taking the drug, they suffer withdrawal symptoms.
- These symptoms may include sweating, shivering, headaches, muscle pain and sickness.

Quick Test
1. Write down a definition for the word 'drug'.
2. Describe what is meant by 'side-effect'.
3. Give two examples of legal recreational drugs.
4. What does the word 'addiction' mean?

Key Words
drug
side-effect
recreational drug
addiction

Our Health and the Effects of Drugs

You must be able to:

- Explain how microbes can cause disease
- Describe how the body acts as a barrier to prevent disease
- Understand how bacteria, viruses and fungi cause disease
- Understand the importance of vaccination and antibiotics.

Microbes

- Microbes or microorganisms are very small organisms that can only be seen by using a microscope.
- Most microbes are harmless to humans but a small number of them can cause disease.

> **Key Point**
>
> Some microbes are even useful to humans, such as yeast used for bread and wine making.

How Microbes Cause Disease

- Microbes can cause disease in one of two ways:
  1. They can attack and destroy cells in our body.
  2. They can produce chemicals called **toxins** that act like poisons in our body.
- Different types of microbes produce different types of diseases in our body.

> **Key Point**
>
> Microbes that cause disease are called **pathogens**.

How the Body Protects us from Disease

- The skin acts as a barrier to stop the microbes entering the body.
- Microbes try to enter the body through body openings and so the body has defenses to stop the microbes getting in. These are shown in the diagram.

How the Blood Protects us Against Microbes

- Sometimes the skin gets damaged and microbes gain entry. We then need a different kind of defence against microbes.
- Our blood can clot to stop microbes from getting into the blood.
- Our blood contains white blood cells. A type of white blood cell called a phagocyte can attack and engulf microbes.

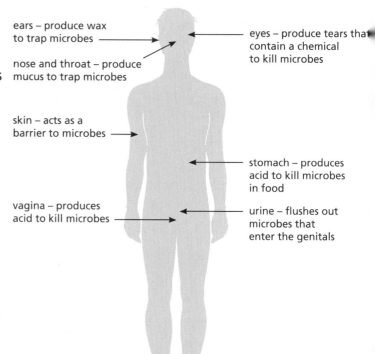

ears – produce wax to trap microbes

eyes – produce tears that contain a chemical to kill microbes

nose and throat – produce mucus to trap microbes

skin – acts as a barrier to microbes

stomach – produces acid to kill microbes in food

vagina – produces acid to kill microbes

urine – flushes out microbes that enter the genitals

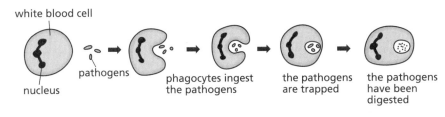

white blood cell

nucleus

pathogens

phagocytes ingest
the pathogens

the pathogens
are trapped

the pathogens
have been
digested

- They can also make microbes clump together and produce chemicals to destroy them.
- There are some white blood cells called **memory cells**. When we get a disease that we recover from, our body makes memory cells. If the same microbe enters the body again the memory cells produce **antibodies** to destroy it even before symptoms develop.

Bacteria, Viruses and Fungi

- There are three different types of microbes that can cause disease, **bacteria**, **viruses** and **fungi**.
- Bacteria can be seen with a light microscope and cause diseases such as tuberculosis.
- Viruses are much smaller and can only be seen with an electron microscope. They cause diseases such as polio.
- Fungi cause diseases such as athlete's foot.

Vaccines and Antibiotics

- Some diseases can be prevented by **vaccination** and some can be cured by **antibiotics**.
- Vaccination is when dead microbes are injected into the body causing the blood to make memory cells. We are then protected against that microbe.
- Antibiotics are chemicals which kill bacteria that have entered our body.

> **Key Point**
>
> White blood cells are part of a defence system called the immune system.

> **Key Point**
>
> Antibiotics do not work against viruses.

> **Key Words**
>
> **toxins**
> **pathogen**
> **memory cell**
> **antibodies**
> **bacteria**
> **virus**
> **fungi**
> **vaccination**
> **antibiotic**

> **Quick Test**
>
> 1. Describe three ways our body stops microbes from entering.
> 2. Name three different types of microbes.
> 3. Explain how vaccinations work.
> 4. Explain why doctors do not prescribe antibiotics for infections caused by a virus.

Getting the Energy your Body Needs

1 Copy the table below and draw a straight line from each description of respiration, to the correct type of respiration.

Type of respiration	Description of respiration	Type of respiration
	Uses oxygen	
	Produces lactic acid	
Aerobic	Produces alcohol	Anaerobic
	Releases the least amount of energy	

[4]

2 Complete the following word equations:

a) for aerobic respiration

   oxygen + _____ → water + _____ + energy   [2]

b) for fermentation in yeast

   glucose → _____ + _____ + energy   [2]

c) for anaerobic respiration in humans

   glucose → _____ + energy   [1]

3 Explain why respiration in living organisms is so important.   [2]

4 The skeleton is an important structure.

Copy the table and put a tick (✓) in the boxes next to each function performed by the skeleton.

Carries oxygen around the body	
Supports the body	
Helps with movement	
Where food is digested	
Protects some organs	
Makes red blood cells	
Where anaerobic respiration takes place	

[4]

5 Joints allow the skeleton to move. Identify the structures numbered 1–5 in the diagram of the knee joint. [5]

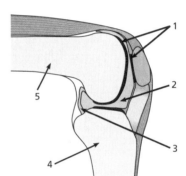

6 Look at the diagram of the human arm.

a) Explain the job done by organ **A**. [2]

b) Explain the job done by organ **B**. [2]

c) What single word best describes these two organs? [1]

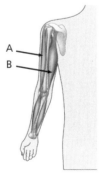

Looking at Plants and Ecosystems

1 a) Complete this word equation for photosynthesis:

water + _____ ➔ glucose + _____ [2]

b) Write down two other things needed for photosynthesis to take place. [2]

2 Using the two gases, carbon dioxide and oxygen, explain how animals and plants are dependent upon each other. [4]

Variation for Survival

1.   Copy and complete the table below by writing down in each box the correct number of chromosomes found in the nucleus of these types of human cell.

Cell	Number of chromosomes
Muscle cell	
Nerve cell	
Sperm cell	
Egg cell	
Embryo cell	

[5]

2.   Give the correct labels for **A** to **D** on the diagram. Choose from the words given below.   [4]

nucleus	cell	gene	chromosome

A

B

C

D

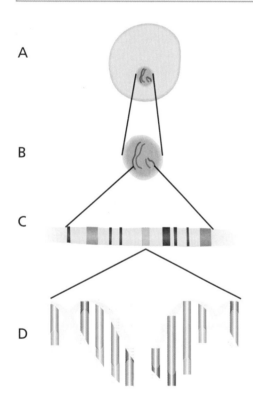

3 The graph shows variation of a characteristic found in humans. Variation can be either continuous or discontinuous.

Use the graph to explain the differences between these two types of variation. [2]

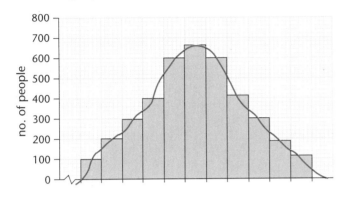

Our Health and the Effects of Drugs

1 Look at the outline drawing of the human body.

Write a description of how each labelled part of the body prevents the entry of microbes. [7]

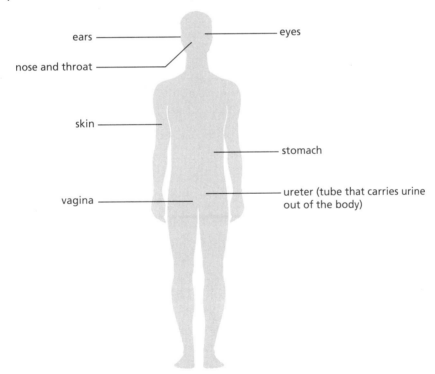

2 Describe the difference between a medical drug and a recreational drug. [2]

3 Recreational drugs can be divided into three categories: depressants, stimulants and hallucinogens. Give one example of each type of drug and describe its effects on the human body. [6]

4 Many people drink alcohol excessively and smoke. Describe the possible long-term effects on the human body of drinking alcohol excessively and smoking. [5]

Mixing, Dissolving and Separating

You must be able to:

- Represent pure substances and mixtures using simple particle pictures
- Demonstrate a range of laboratory skills
- Apply appropriate separation techniques to different mixtures.

Quick Recall Quiz

Pure and Impure Substances

- In chemistry a 'pure' substance is one that contains only one type of atom or compound.
- An impure substance contains more than one substance (element or compound), forming a mixture.
- The substances in the mixture are not chemically joined together so it should be easy to separate them.
- A common example of a mixture is sugar dissolved in water.

Particle diagram of sugar dissolved in water

Water molecules in a pure solution

A sugar solution with water and sugar molecules

Chromatography

- Chromatography separates dissolved pigments in solution, e.g. the pigments in ink:
  1. A drop of ink is placed on a pre-marked line at the bottom of a piece of chromatography paper and the paper is dipped into the **solvent**.
  2. As the solvent moves up the paper it takes the dissolved pigments with it.
  3. Since the pigments have different solubilities they travel at different speeds and so separate. Insoluble substances do not move up the paper.
  4. The most soluble pigments move the furthest; less soluble pigments move less far. Pure substances produce only one spot of colour.

Chromatography

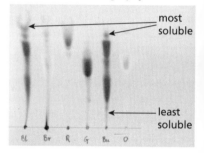

most soluble

least soluble

Key Point

A pencil line marks the starting point as it will not move with the ink pigments.

Filtering

- **Filtration** separates an insoluble solid from a liquid or solution by passing the solid/liquid mixture through filter paper.
- The **filtrate** is the liquid or solution which passes through the filter paper and the solid left behind is the residue.
- Excess copper oxide in copper sulfate solution can be separated by filtration. Copper sulfate solution is the filtrate; copper oxide is the residue.

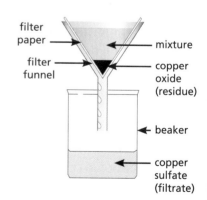

filter paper

filter funnel

mixture

copper oxide (residue)

beaker

copper sulfate (filtrate)

Evaporation

- **Evaporation** is used to remove the liquid part of a mixture and collect the dissolved solid.
- The mixture is placed in a suitable container (e.g. a watch glass) and heated, sometimes by using a Bunsen burner. This is called crystallisation.

Distillation

- Liquids have different boiling points.
- By carefully controlling the temperature of a heated mixture of two or more liquids, the liquids evaporate at different times. This is known as distillation.
- The evaporated gas is cooled and condenses into a liquid and collected as **distillate** in a collecting vessel.
- The fragrances used in perfumes are separated by distillation, as well as the different parts of crude oil.

Key Point

The slower the liquid evaporates, the larger the crystals that form.

Key Point

Pure substances have a melting point and a boiling point. Mixtures melt and boil over a range of temperatures.

Distillation of ethanol (an alcohol) and water mixture

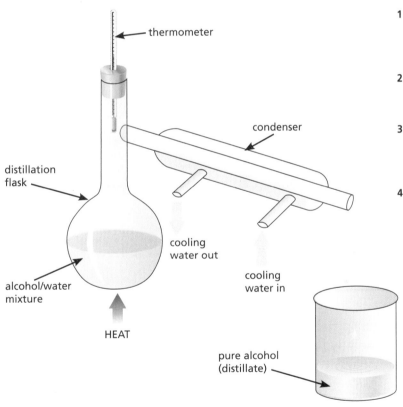

- thermometer
- condenser
- distillation flask
- cooling water out
- alcohol/water mixture
- HEAT
- cooling water in
- pure alcohol (distillate)

1 The mixture is heated until the liquid with the lowest boiling point (ethanol) boils.

2 The thermometer measures the temperature of the gas.

3 The water in the condenser cools the gas, allowing it to condense back into a liquid.

4 The liquid (distillate) is collected.

Quick Test

1. How could you separate an insoluble solid from a liquid?
2. What size crystals are made from rapid evaporation?
3. What does 'distillate' mean?
4. Describe how to carry out a chromatography experiment.

Key Words

solvent
filtration
filtrate
evaporation
distillate

Mixing, Dissolving and Separating

You must be able to:

- Explain the conservation of mass in reactions and changes of state
- Represent pure substances and mixtures using particle pictures and word equations
- Explain similarities and differences between combustion, thermal decomposition, oxidation and reduction.

Conservation of Mass

- The law of conservation of mass states that in any physical change or chemical reaction the total mass after the change will be the same as the total mass before the change.
- With state changes this means that the number of particles of the substance at the start will equal the number of particles at the end.

Conservation of mass

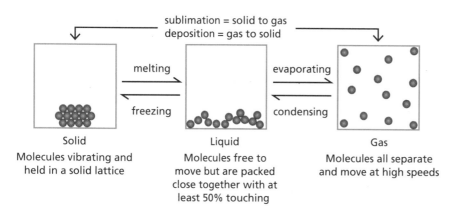

sublimation = solid to gas
deposition = gas to solid

melting / freezing

evaporating / condensing

Solid
Molecules vibrating and held in a solid lattice

Liquid
Molecules free to move but are packed close together with at least 50% touching

Gas
Molecules all separate and move at high speeds

> **Key Point**
>
> When drawing a particle diagram for a liquid, at least half of the particles should be touching each other.

- With a chemical reaction the atoms of the reactants are rearranged to form the products. Atoms cannot 'disappear'.

Combustion

- **Combustion** is the reaction between a fuel and oxygen.
- Carbon dioxide and water are generally produced as waste products when the fuel is a hydrocarbon.
- Energy moves from the chemical store to the thermal store.

fuel + oxygen ⟶ carbon dioxide + water + energy

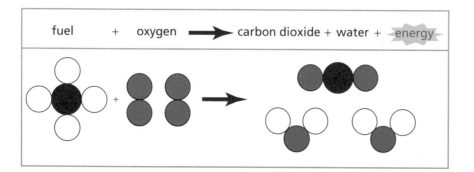

Thermal Decomposition

- Some compounds break down into new molecules when heated; they don't react with oxygen in the air.
- This is called **thermal decomposition**.
- An example is chalk, which has the chemical name calcium carbonate.

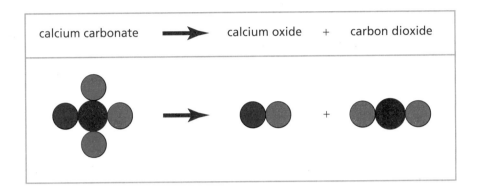

calcium carbonate ➡ calcium oxide + carbon dioxide

Key Point

In this model, the coloured circles represent atoms. There are the same number on each side of the equation.

Oxidation and Reduction

- When substances gain oxygen in a reaction it is called **oxidation**.
- Losing oxygen in a reaction is called **reduction**.
- For example, carbon can be oxidized to form carbon dioxide:

Oxidation forming rust

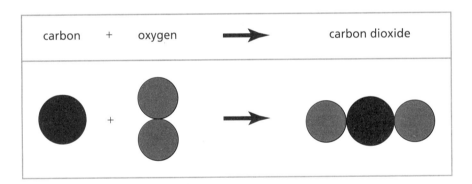

carbon + oxygen ➡ carbon dioxide

- The reaction of iron with water and oxygen is a special form of oxidation, forming iron(III) oxide, which is known as **rust**.

iron + water + oxygen ➡ **hydrated iron(III) oxide**

- Rusting requires oxygen and water. It happens faster when salt is dissolved in the water.

Key Point

Only use the term rust for the oxidation of iron. Other metals corrode, they don't rust.

Key Words

combustion
thermal decomposition
oxidation
reduction
rust

Quick Test

1. Draw diagrams to show the atoms in a solid, liquid and gas.
2. Describe what happens when calcium carbonate is heated.
3. What is meant by the term oxidation?
4. What does iron have to react with in order to rust?

Elements, Compounds and Reactions

Quick Recall Quiz

You must be able to:

- Explain the structure of the periodic table, including groups, periods, symbols and formulae
- Explain differences between elements and compounds in terms of particles
- Describe the differences between physical changes and chemical reactions.

The Periodic Table

- The periodic table contains all the **elements** that are found in the universe.
- An element is a substance that contains only one type of atom.
- The simplest particle of an element that cannot be broken down further without losing its properties is called an **atom**.
- The periodic table arranges the elements based on the atomic number (the number of protons in the nucleus of each atom) and the physical and chemical properties of each element.
- Each column of the periodic table is called a **group**, a family of elements with similar physical and chemical properties.
- The rows in the periodic table are called **periods**. The atomic number increases from left to right through the period.
- The majority of elements in the periodic table are metals; the non-metals are less common.

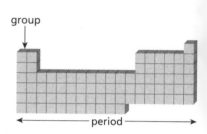

group

period

> ### Key Point
>
> The atomic number of elements increases sequentially in whole numbers as you go through the table.

Key

Metals

Non-metals

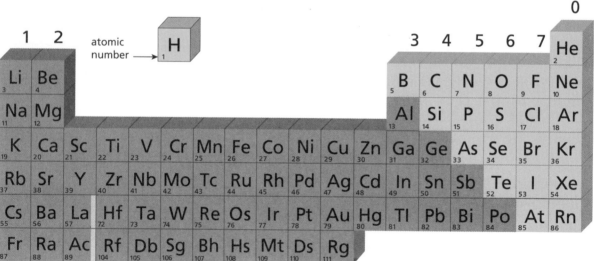

- The periodic table used today was devised by the Russian chemist Dmitri Mendeleev.
- We can use the periodic table to predict the physical and chemical properties of elements.

For example, elements in a group are very similar. They have similar physical properties and chemical reactions:
- The metals in group 1 all react with water to form alkaline solutions
- The non-metals in group 7 are good at killing bacteria.

Chemical Symbols and Formulae

- Elements have a name and a chemical symbol.
- Symbols consist of one or two letters, for example Helium = He, Copper = Cu, Silver = Ag.
- When chemicals react and chemically join together they form compounds.
- The compound is represented by a chemical formula, e.g. water = H_2O, where two hydrogen atoms are joined to one oxygen atom.
- The number written as a subscript indicates how many of those atoms are in the compound, e.g. $C_6H_{12}O_6$ (glucose) contains 6 carbon, 12 hydrogen and 6 oxygen atoms.

Physical Changes and Chemical Reactions

- A physical change is where a substance changes state, e.g. water (liquid) freezing into ice (solid).
- Physical changes are easy to reverse.
- A chemical reaction is where elements chemically join together to form a compound.
- It is difficult to reverse a chemical reaction.
- The compound formed has different properties to those of each of the original elements.
- To make it clear what is happening in a reaction we write a chemical equation:
    - On the left hand side we write the **reactants**
    - On the right hand side we write the **products** formed.
- An example is the reaction of sodium with chlorine:

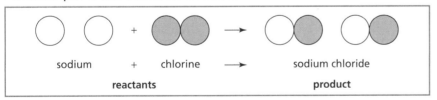

sodium	+	chlorine	⟶	sodium chloride
	reactants			**product**

Quick Test

1. Which of the following are compounds?
   O_2 CO_2 H_2O
2. What is the atomic number of an element?
3. How many hydrogen atoms are in H_2SO_4?
4. How many different elements make up $C_6H_{12}O_6$?

Elements, Compounds and Reactions

Quick Recall Qui

You must be able to:

- Describe the properties of metals and non-metals
- Explain the reactions of metals and metal oxides with acids
- Understand the concept of a reaction using oxidation of metals and non-metals.

Properties of Metals and Non-metals

Metal – copper wires used to conduct electricity

Metals	Non-metals
Conduct electricity and heat	Are unable to conduct electricity and heat
Are **ductile** (can be drawn into wires)	Often have a low melting point and boiling point
Are **malleable** (can be hammered into shape)	Are often gases at room temperature
Are shiny	Often have a lower **density** than metals.
Are sonorous (ring like a bell when hit)	
Often have a high melting point and boiling point.	

Non-metal – helium gas used to inflate balloons

Reactions of Metals

- Metals react with acids to give a **salt** and hydrogen:

$$\text{metal} + \text{acid} \longrightarrow \text{salt} + \text{hydrogen}$$

- The salt formed always takes the name of the metal plus a suffix that represents the acid used in the reaction:

Key Point

Hydrogen is a gas, so bubbles are always produced when acid and metal react.

Acid	Suffix	Example
Hydrochloric acid	Chloride	**magnesium + hydrochloric acid ⟶ magnesium chloride + hydrogen**
Sulfuric acid	Sulfate	**magnesium + sulfuric acid ⟶ magnesium sulfate + hydrogen**
Nitric acid	Nitrate	**magnesium + nitric acid ⟶ magnesium nitrate + hydrogen**
Phosphoric acid	Phosphate	**magnesium + phosphoric acid ⟶ magnesium phosphate + hydrogen**

Oxidation

- Reacting an element or compound with oxygen is called **oxidation**.
- The atoms that make up the elements and compounds rearrange to make a new compound, an oxide. For example:

> magnesium + oxygen ⟶ magnesium oxide

- This can also be written as a balanced equation.
- A balanced equation indicates the number of atoms and how they are arranged.
- For example, two atoms of magnesium react with one molecule of oxygen to form two lots of magnesium oxide:

> $2Mg(s) + O_2(g) \longrightarrow 2MgO(s)$

- Combustion is where a fuel reacts with oxygen (burns) forming carbon dioxide and water, and giving out energy in the process:

> fuel + oxygen ⟶ carbon dioxide + water + energy

Reactions of Metal Oxides

- Metals react with oxygen to form metal oxides, for example:

> zinc + oxygen ⟶ zinc oxide

- The metal oxide is called a **base** and is the chemical opposite of an acid.
- Metal oxides react with acids to form a salt and water:

> metal oxide + acid ⟶ salt + water

- This means that the acid has been **neutralised**.
- The salt formed always takes the name of the metal and the suffix from the acid, for example:

> copper oxide + hydrochloric acid ⟶ copper chloride + water

> **Key Point**
>
> The numbers of each atom are always the same on both sides of the equation.

> **Key Point**
>
> Metal oxides dissolve to form alkaline solutions. Non-metal oxides dissolve to form acidic solutions.

Sodium Chloride is one type of salt

> **Key Words**
>
> ductile
> malleable
> density
> salt
> oxidation
> base
> neutralise

Quick Test

1. Write the word equation for the reaction between calcium and oxygen.
2. What salt is formed in the reaction between zinc and sulfuric acid?
3. Give three properties of a metal.
4. Give three properties of a non-metal.

Variation for Survival

1 The graph shows variation of a characteristic found in humans.

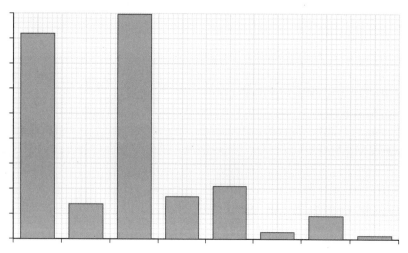

a) State the name of the type of variation shown in the graph. [1]

b) Write down two examples of this type of variation found in humans. [2]

2 Variation is very important to the survival of a species.

Under which of the following conditions is variation most important? Put a tick next to the best answer from the table below.

When environmental conditions stay the same	
When the environment is changing very slowly	
When the environment is changing very quickly	
The environment has no effect on variation	

[1]

3 Explain the part played by each of the following scientists in the understanding of DNA.

a) Watson and Crick. [1]

b) Rosalind Franklin. [1]

Our Health and the Effects of Drugs

1 Define the following scientific words.

   a) pathogen

   b) toxin

   c) antibody   [3]

2 Name three different types of microbe and give an example of a disease caused by each of them. [3]

3 Explain why doctors are less worried about people taking the drug caffeine and more worried about people taking the drug cocaine. [2]

4 Describe the difference between addiction and withdrawal. [2]

5 Match each activity below with a danger of doing it.

Activity	Danger
Drinking alcohol	Hallucinations
Smoking	Reduces breathing
Taking LSD	Lung cancer
Using heroin	Liver failure

   [4]

6 Microbes sometimes gain entry to our body.

Describe how white blood cells help the body fight infection. [3]

Mixing, Dissolving and Separating

1 Which of the following separation methods would be best for each of the following investigations?

Choose from the following methods:

chromatography   **distillation**   **filtration**

a) Extracting alcohol from beer. [1]

b) Checking whether a note written in blue ink was written using a particular pen. [1]

c) Separating sand from a mixture of sand and salty water. [1]

2 a) Which two of the following diagrams show a pure substance at room temperature? [2]

    A   B   C   D

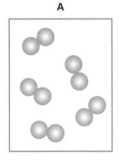 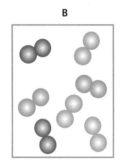 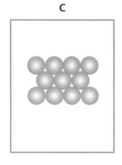 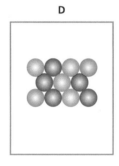

b) Which of the substances could be copper? [1]

3 Describe how filtration and evaporation could be used to extract salt from muddy seawater. [5]

4 Vinnie is analysing the pigments used to colour different sweets using chromatography.

Vinnie makes a qualitative observation on how similar the chromatograms are to each other using his eyes and judgement.

Suggest what Vinnie would need to do to make a more accurate quantitative measurement. [1]

5 An aquarium for keeping fish uses a filter.

Suggest what the filter is removing from the water and explain how filtration works. [3]

Elements, Compounds and Reactions

1   Ethan is painting a model. When he opens a tin of paint he notices that the paint has separated into layers.

    a) What type of a substance is the paint?

      i) an element   **ii)** a compound   **iii)** a mixture   **[1]**

    b) When Ethan reads the label on the tin it says that the paint contains water and titanium oxide. Titanium oxide and water are examples of what type of substance?

      i) elements   **ii)** compounds   **iii)** mixtures   **[1]**

2   Which of the following are examples of chemical reactions?

    i) chocolate melting on a hot day

    ii) a firework exploding in the sky

    iii) bread being toasted

    iv) water condensing on a cold window pane   **[2]**

3   A periodic table is shown below.

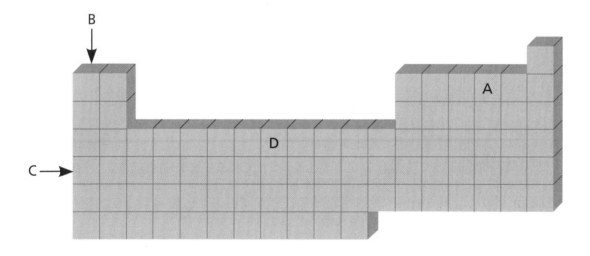

    a) Which letter indicates a group?   **[1]**

    b) Which letter indicates a period?   **[1]**

    c) Which letter indicates a non-metal?   **[1]**

4   Explain the difference between an element and a compound.   **[3]**

Explaining Physical Changes

You must be able to:

- Describe the similarities and differences between solids, liquids and gases
- Explain how changes in temperature affect the motion and spacing of particles
- Explain sublimation in terms of particles.

The Particulate Nature of Matter

- All matter in the universe is made up of **atoms**, arranged in one of three states: solid, liquid or gas.
- At the coldest temperature possible (–273 °C or 0 K), the atoms have no **kinetic energy** so cannot move.
- If heat is introduced, the atoms gain kinetic energy and so move.

Solid

Solids, Liquids and Gases

- Solids:
  - contain atoms arranged as close together as possible
  - are therefore denser than their liquid form (apart from water) and cannot be compressed
  - will have a fixed shape and volume that does not depend upon the container that it is in.
- Even though they form part of a solid the atoms, or molecules, still vibrate due to their kinetic energy.
- As the temperature supplied to a substance increases, the atoms or molecules vibrate more and more.
- Eventually, at the melting point, the atoms or molecules rearrange into a liquid:
  - as the atoms or molecules are slightly further apart the **density** will be less than it was as a solid
  - if a liquid is in a container it will take the shape of the container that it occupies
  - the atoms or molecules move around much more than in a solid, but still cannot be compressed.
- Eventually, at the boiling point, the liquid becomes a gas:
  - the atoms or molecules in a gas can move freely and will occupy all of the available space in a container
  - if not completely enclosed, the gas particles will escape
  - when a material cools, the reverse process happens.
- Unlike solids and liquids, gases can be compressed.
- If kinetic energy is removed from the substance, the particles move more slowly.
- The substance changes from a gas into a liquid (condenses), then into a solid from a liquid (freezing).

Liquid

Gas

> ### Key Point
>
> Gases can still be heated further, to temperatures higher than the boiling point.

Solids	Liquids	Gases
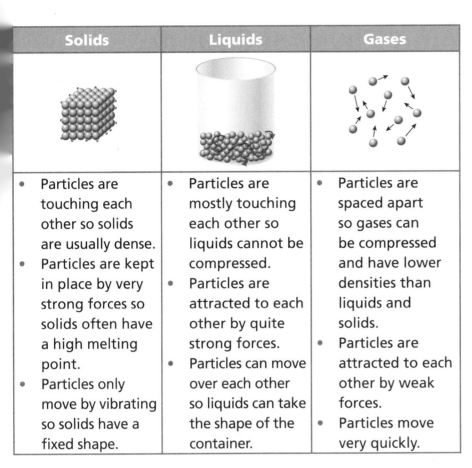		
• Particles are touching each other so solids are usually dense. • Particles are kept in place by very strong forces so solids often have a high melting point. • Particles only move by vibrating so solids have a fixed shape.	• Particles are mostly touching each other so liquids cannot be compressed. • Particles are attracted to each other by quite strong forces. • Particles can move over each other so liquids can take the shape of the container.	• Particles are spaced apart so gases can be compressed and have lower densities than liquids and solids. • Particles are attracted to each other by weak forces. • Particles move very quickly.

Sublimation

- Some substances can jump from solid to gas. This is called **sublimation**.
- Examples of sublimation include carbon dioxide (dry ice to gas) and gel air freshener.

Gel air freshener

> **Key Point**
>
> Sublimation is the change from a solid direct to a gas.

Quick Test

1. Why are solids denser than their gaseous form?
2. At what temperature do atoms stop vibrating?
3. Describe what happens in sublimation.
4. State the differences between solids, liquids and gases, in terms of the arrangement and movement of particles.

Key Words

atom
kinetic energy
density
sublimation

Explaining Physical Changes

You must be able to:

- Describe the particular nature of water and the ice/water transition
- Describe and explain Brownian motion and the diffusion of gases
- Explain the process of heat conduction between particles in a conductor.

Quick Recall Quiz

Water

- Water has a number of properties that are unique.
- When ice forms, the water molecules line up in a regular pattern.
- The water molecules are further apart in ice than in the liquid form, and therefore solid water is less dense than liquid water.
- Consequently, ice floats on water.

Brownian Motion

- In the 1800s Robert Brown observed pollen grains suspended in water under the microscope.
- He noticed that the particles were moving randomly in the water, and his observation is now called **Brownian motion**.
- Brownian motion is due to the particles suspended in a fluid colliding with the atoms or molecules that make up the fluid.

> **Key Point**
>
> Brownian motion is the random movement of tiny particles caused by collisions with the other moving particles in the liquid or gas.

Diffusion

- **Diffusion** is the name of the process whereby molecules in a liquid or gas mix as a result of their random motion.
- Particles at a high **concentration** in one location will tend to move to an area where they are in low concentration.
- Eventually the particles will become evenly distributed throughout the liquid or gas.

Conduction

- When a solid conductor, such as metal, is heated, the atoms increase their energy and vibrate more.
- The atoms collide with other atoms, transferring energy and causing them to vibrate more.
- The process of conduction continues until all the atoms have reached the same temperature.
- In an insulator, the energy is not passed onto other atoms so the solid does not conduct the heat.
- The vibration of the particles increases as their kinetic energy increases.
- The higher the temperature, the greater the kinetic energy and so the particles vibrate faster.
- The lower the temperature, the lower the kinetic energy and the particles will vibrate more slowly.

Temperature and Particles

- The hotter particles get, the more kinetic energy they have.
- This means particles move more and separate from each other more.
- As temperature increases, **pressure** will increase and the density will decrease.
- In the case of a balloon, the particles inside will increase in speed causing the pressure to increase, enlarging the balloon.

Movement of particles inside a warmed balloon

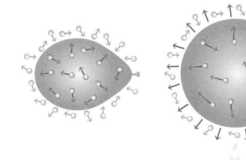

gas pressure depends on both density and temperature

heating the air also increases the pressure

- In exothermic reactions, the temperature increases as energy moves from the chemical store to the thermal store.
- In endothermic reactions, the temperature decreases as energy moves from the thermal store to the chemical store.

Quick Test

1. What will happen to the air particles in a sealed balloon if it is heated?
2. What will happen to the air particles in a sealed balloon if it is cooled?
3. Why does ice float on water?
4. Explain how diffusion takes place.

Key Words

Brownian motion
diffusion
concentration
pressure

Explaining Chemical Changes

Quick Recall Quiz

You must be able to:

- Apply conservation of mass to simple reactions
- Explain the combustion of fuels
- Explain the difference between a chemical and physical change
- Explain how a catalyst can make a reaction occur faster by reducing activation energy.

Chemical Reactions

- A chemical reaction involves the rearrangement of atoms from reactants to products.
- The products that are formed have the same atoms, just in different configurations.
- There is never a change in total mass in a chemical reaction.
- A **word equation** names the reactants and products formed in a reaction, for example:

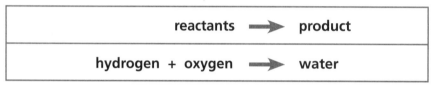

reactants ➡ product
hydrogen + oxygen ➡ water

> **Key Point**
>
> The total mass in a chemical reaction doesn't change. This is conservation of mass.

- Word equations do not tell us the ratios of the molecules involved.
- **Chemical equations** show the chemical formula of the reactants and products, so that the number of atoms and ratios involved can be worked out, for example:

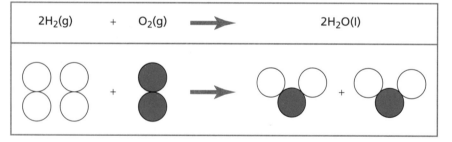

$$2H_2(g) \quad + \quad O_2(g) \longrightarrow 2H_2O(l)$$

> **Key Point**
>
> The large number in front of each molecule or atom is a coefficient. It tells us how many molecules or atoms there are overall.

- This equation tells us that two molecules of hydrogen gas react with one molecule of oxygen gas to give two molecules of water.
- The state of each reactant is given in brackets after its chemical formula: (s) = solid, (g) = gas, (l) = liquid, (aq) = aqueous (which means it is dissolved in water).
- In combustion, a fuel generally reacts with oxygen to produce carbon dioxide and water, and releases energy in the process. For example, burning propane gas in a camping stove:

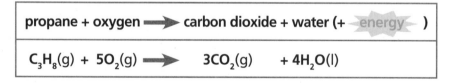

propane + oxygen ➡ carbon dioxide + water (+ energy)
$C_3H_8(g) + 5O_2(g) \longrightarrow 3CO_2(g) + 4H_2O(l)$

Catalysts

- A chemical reaction will only take place if the particles have enough energy to react.
- This is called the **activation energy** of the reaction.
- If the particles have less energy than the activation energy, there will not be a reaction.
- A **catalyst** is a substance that reduces the activation energy, and so increases the rate of reaction.
- This means the reaction can take place with lower energy than normal, so more particles can react.
- A catalyst is neither a reactant nor a product and is not used up in the reaction.
- The name of the catalyst is written above the arrow in the reaction to indicate that it is needed.
- Catalysts are used in chemical processes all over the world, most commonly in the exhaust systems of cars.

> **Key Point**

Catalysts are never used up, so can be used again and again.

Catalytic convertor

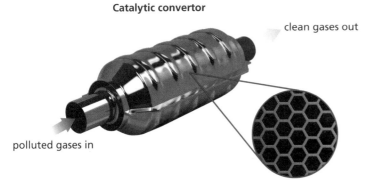

clean gases out

polluted gases in

Pt catalyst to catalyse
the breakdown of pollutants

| **Pt catalyst** |
| carbon monoxide + oxygen ➡ carbon dioxide |
| **Pt catalyst** |
| 2CO(g)   +   O$_2$(g) ➡   2CO$_2$(g) |

- In biology, catalysts are made of protein and are called enzymes.
- One example of an enzyme is amylase in the digestion of starch into the sugar maltose:

amylase

starch ➡ maltose

> **Quick Test**

1. What does a catalyst do?
2. Explain the term activation energy.
3. State the differences between word and chemical equations.
4. Write the four state symbols used in equations.

> **Key Words**

word equation
chemical equation
activation energy
catalyst

Explaining Chemical Changes

You must be able to:

- Explain neutralisation and the use of indicators
- Use word equations to represent and/or describe the reactions of acids
- Describe and explain the uses of acids and alkalis.

Quick Recall Quiz

Indicators

- Indicators are used to find out what type of substance a chemical is.
- Initially, scientists discovered chemicals in common plants that could change colour to indicate acid or alkali, for example, red cabbage is red in acid and blue in alkaline conditions.
- The chemical litmus also behaves as an indicator but can be incorporated into paper, so it can be transported easily.
- Universal indicator (UI) solution and paper contain a mixture of different indicators.
- These indicators change colour at a specific pH.
- pH is a measure of the strength of an acid or alkali.
- The pH scale ranges from 1 to 14 and has a colour for each pH number.

Litmus paper

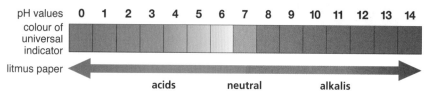

- pH probes and data loggers remove the need for indicator papers and solutions.
- They measure the pH directly and are more precise than indicator papers or solutions.

Acids and Bases

- All chemicals can be classified as being an **acid**, a **base** or are **neutral**.
- Acids are a group of chemicals that have a pH less than 7.
- An acid can chemically react with a metal to produce hydrogen:

metal + acid ⟶ salt + hydrogen

- The chemical opposite of an acid is a base.
- Bases are chemicals with a pH greater than 7.

> ### Key Point
>
> If a base dissolves, it's an **alkali**.

Name of acid	Where found	pH
Hydrochloric acid	Human stomach	1
Ethanoic acid	Vinegar	2
Citric acid	Citrus fruit	2
Sulfuric acid	Car batteries	1
Carbonic acid	Fizzy drinks	4

Name of base	Where found	pH
Sodium hydroxide	Laboratories	14
Calcium carbonate	Chalk	9
Sodium bicarbonate	Bicarbonate of soda (cooking)	8
Ammonia	Hair dyes	11
Lime	Gardening products	12

Citrus fruit – acid

Chalk – base

- When a chemical is neither an acid nor a base it is neutral.
- A neutral solution has a pH of 7.

Neutralisation

- When an acid and a base are mixed together, they react.
- If an acid is reacted with a base there will come a point where a salt and water are made and no more acid or base exists.
- At this point the mixture will be neutral and have a pH of 7.
- The whole process is called **neutralisation**.

acid + metal oxide $\longrightarrow$ salt + water
acid + metal hydroxide $\longrightarrow$ salt + water
acid + metal carbonates $\longrightarrow$ salt + water + carbon dioxide

Key Point

In all neutralisation reactions, water is made.

Quick Test

1. What can you deduce if a chemical is pH 5?
2. Explain neutralisation.
3. Write the general equation for the reaction of acid and metal.
4. Why is a pH probe and a data logger better to measure pH than UI paper?

Key Words

acid
base
neutral
alkali
neutralisation

Mixing, Dissolving and Separating

1 Anna and Kala are carrying out chromatography of ink, as they believe a cheque has been forged. They are going to test to see whether the ink on the cheque is the same as the ink of the suspect's pen. The results are shown in the chromatogram below.

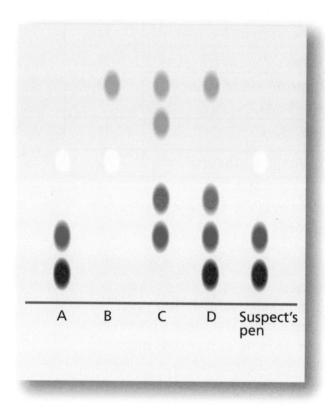

a) How many different pigments are in sample A and in sample B? [2]

b) Which of the samples matches the ink of the suspect's pen? [1]

c) Explain why was the starting line on the chromatogram was drawn using a pencil. [2]

2 Karim is investigating how chalk (calcium carbonate) reacts when heated. He heats the chalk for a minute at a time and then measures its mass.

He notices that the mass decreases.

a) What is the name given to the type of chemical reaction Karim is observing? [1]

b) Complete the word equation for the reaction given below:

calcium carbonate ➔ _____ + _____ [2]

3 Write the chemical equation for the reaction of carbon with oxygen to form carbon dioxide. Include state symbols. [3]

Elements, Compounds and Reactions

1 The diagram shows a model of a chemical reaction.

a) What feature of the diagram indicates that a chemical reaction has taken place?   [2]

b) Substance X is oxygen. Suggest what substances Y and Z could be.   [2]

c) Suggest how the diagram indicates that mass has been conserved in the reaction.   [1]

   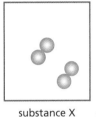   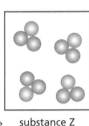

substance X   +   substance Y   →   substance Z

2 Niamh is carrying out some experiments to identify what happens when different metals are added to different acids.

Her table has some gaps. What are the reactants and products given by letters A–E?

Metal	Name of acid	Salt formed	Gas formed
Zinc	A	Zinc sulfate	Hydrogen
Magnesium	Nitric acid	B	Hydrogen
Iron	C	Iron chloride	D
E	Nitric acid	Lead nitrate	Hydrogen

[4]

3 For each change given below, decide whether it is a physical change (P) or a chemical change (C).

a) melting chocolate

b) a burning firework

c) the smell of perfume diffusing across a room

d) jelly setting in a dish

e) an iron nail rusting   [1]

4 Which of the following chemical equations is correct?

a) $Mg(s) + O_2(g) \rightarrow 2MgO(s)$

b) $2Mg(s) + O(g) \rightarrow MgO(s)$

c) $2Mg(s) + O_2(g) \rightarrow 2MgO(s)$

d) $Mg(s) + O(g) \rightarrow MgO(s)$ [1]

Explaining Physical Changes

1 Draw the particles in a solid, liquid and a gas. [3

2 Sally is looking at a blue coloured gel air freshener, which sublimes. She cuts a piece of the air freshener and puts it into a small beaker. The beaker is placed into a larger beaker which contains hot water. On the top of the beaker containing air freshener, she places a beaker containing ice.

    a) What would Sally see after a few minutes on the underside of the beaker containing ice? [1]

    b) Suggest why this happens. [1]

3 What is meant by the term Brownian motion? [2]

4 An iceberg floats on water. Explain why ice floats on liquid water. [3]

5 Claude is holding a CO_2 fire extinguisher. When he sets the extinguisher off, a valve opens releasing CO_2 gas.

When a handkerchief is placed over the end of the extinguisher, solid CO_2 starts to build up.

Which of the following best explains why this happens? Tick the correct box.

    a) The CO_2 particles break up into C and O_2 ☐

    b) The CO_2 gas particles are slowed down rapidly causing a solid to form ☐

    c) The CO_2 gas particles slow down to form a liquid and then a solid ☐

    d) The handkerchief cools the CO_2 down into a solid ☐ [1]

Explaining Chemical Changes

1 Mark is reacting copper metal with oxygen gas. He draws the atoms involved in the reaction.

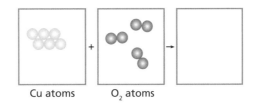

Cu atoms O_2 atoms

a) Draw what should be in the last box. [2]

b) What is the name of the product in this reaction? [1]

2 What does a catalyst do in a chemical reaction? Choose the best answer.

a) It slows a chemical reaction down

b) It makes chemical reactions happen

c) It lowers the activation energy so the reaction happens faster

d) It lowers the kinetic energy [1]

3 Complete the following:

metal + acid ➜ .... + ... [1]

4 Peter is using universal indicator to identify acids and bases.

Suggest what colour UI would change to for each of the examples below:

a) car battery acid

b) juice of a lemon

c) tap water

d) toothpaste [4]

5 What are the products of the following reactions?

a) magnesium + hydrochloric acid ➜ [1]

b) copper oxide + nitric acid ➜ [1]

c) vanadium carbonate + sulfuric acid ➜ [1]

Obtaining Useful Materials

You must be able to:

- Use the reactivity series to determine whether reactions are possible
- Describe and explain how carbon is used to extract metals.

The Reactivity Series

- The metals in the periodic table all exhibit different levels of **reactivity**.
- By comparing their reactions it is possible to sort them into a reactivity series.
- The reactivity series shows the order of reactivity, from most reactive to least reactive.

> **Key Point**
>
> All metals could be included in a reactivity series. The diagram here shows only a few major metals, as well as carbon.

Most reactive

Potassium

Sodium

Calcium

Magnesium

Aluminium

Carbon

Zinc

Iron

Tin

Lead

Copper

Silver

Gold

Platinum

Least reactive

Structure of diamond
Diamond is very strong

- By comparing metals on the reactivity series, chemists can predict whether a chemical reaction may happen and, if it does, how vigorous the reaction may be.
- The reactivity series also includes the element carbon.
- Carbon can have different forms and properties, such as diamond and graphite.
- Carbon is not a metal, but when it is in the form of graphite it can conduct electricity, a property shared with the metals.

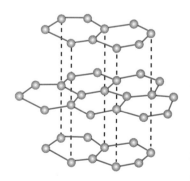

Structure of graphite
Graphite conducts electricity

Displacement Reactions

When a metal, or carbon, comes into contact with a metal that is in a compound there may be a reaction:

- If the metal in the compound is higher in the reactivity series than the introduced metal, or carbon, no reaction will take place.
- If the metal in the compound is lower in the reactivity series than the introduced metal or carbon then the introduced metal will **displace** the metal in the compound.
- The greater the difference in positions in the reactivity series, the faster and more vigorous the reaction.

For example, if iron metal was added to copper sulfate solution:

> **Key Point**
>
> Remember that, as well as new products being formed, there will often be visible changes as a result of the reactants disappearing.

Iron nail in copper sulfate solution

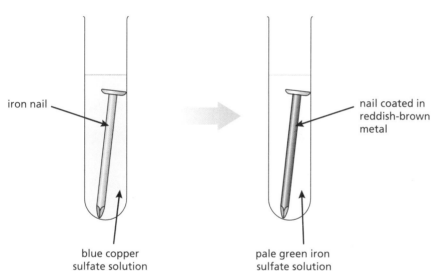

iron nail

nail coated in reddish-brown metal

blue copper sulfate solution

pale green iron sulfate solution

Iron is higher than copper on the reactivity series. It will therefore displace the copper and form iron sulfate. Iron is a silver coloured metal. The blue copper sulfate solution will lose its colour as the copper is displaced and pale green iron sulfate is formed. The copper, which was previously in solution, will form a reddish-brown solid.

- The reaction can be written as:

iron + copper sulfate $\longrightarrow$ iron sulfate + copper
$Fe(s) + CuSO_4(aq) \longrightarrow FeSO_4(aq) + Cu(s)$

- If copper were added to iron sulfate solution, there would be no reaction as copper is lower in the reactivity series than iron and will not be able to displace it.

> **Quick Test**
>
> 1. Could copper displace magnesium from magnesium nitrate?
> 2. Write a word equation for the reaction between magnesium and iron sulfate.
> 3. Suggest a metal to displace sodium from sodium chloride.
> 4. Explain what is meant by the 'reactivity series'.

> **Key Words**
>
> reactivity
> displace

Obtaining Useful Materials

You must be able to:

- Describe and explain the use of carbon in the extraction of metals
- Describe the characteristics and uses of ceramics, composites and polymers.

Extracting Metals

- Many of the first metals discovered by humans were low on the reactivity series.
- As they were unreactive it meant that the metals could be found in their pure form, e.g. gold and silver, and were not in a compound.
- Other, more reactive, metals tended to have already reacted with other elements, such as oxygen, to form compounds.
- One of the biggest milestones in human history was stumbling upon the displacement reaction using carbon that can be used to purify iron ore:
  - Iron ore is heated in a furnace with carbon and limestone
  - Carbon is higher in the reactivity series than iron so displaces the iron
  - Molten iron is formed and the carbon joins on to the oxygen forming carbon dioxide.

> **Key Point**
>
> All metals below carbon in the reactivity series can be extracted using this method.

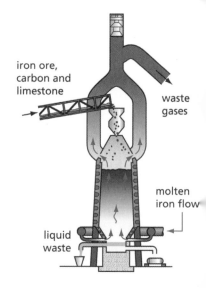

iron ore, carbon and limestone

waste gases

molten iron flow

liquid waste

carbon + iron(III) oxide ⟶ carbon dioxide + iron
$3C(s) + 2Fe_2O_3(s) \longrightarrow 3CO_2(g) + 4Fe(l)$

Ceramics

- Ceramics are made from heating non-metallic materials at high temperatures.
- The properties of the **ceramic** material differ from the initial material, e.g. ceramic pots are very different to the clay used to make them.
- This is because the high temperatures cause crystals to form on cooling.
- By controlling the speed of cooling, different sized crystals can be made.
- Rapid cooling causes small crystals and slow cooling causes large crystals.

- If other minerals are added to glass when it is formed, glass ceramics can be made.
- Although they are often brittle, these can tolerate very high temperatures, so are often used as cookware (e.g. Pyrex) or in laboratories.

Polymers

- It is possible to join small molecules (**monomers**) together in long chains.
- The chained molecule consisting of repeating monomer units is called a **polymer**.
- Polymers are extremely useful because they can be used to create different materials, e.g. the monomer ethene can be made into poly(ethene).
- Poly(ethene) is used to make plastic bags and bottles.

Key Point

Materials such as clay need to be heated in an oven to become a ceramic. This is an irreversible process.

$$n \quad \begin{array}{c} H \quad H \\ | \quad | \\ C = C \\ | \quad | \\ H \quad H \end{array} \rightarrow \left(\begin{array}{c} H \quad H \\ | \quad | \\ C - C \\ | \quad | \\ H \quad H \end{array} \right)_n$$

ethene   poly(ethene)

n means a large number of, or many

Composites

- A **composite** is a material that is made from two or more different materials bonded (joined) together.
- The new composite material has different characteristics to those of the starting materials.
- Concrete, first made by the Romans, is made from mixing cement with different stones. The resulting mixture is far stronger than the cement or the stones alone.
- Carbon fibre is a very light and exceptionally strong composite formed from sheets of carbon fibre bonded together with a resin.
- Carbon fibre materials are used when strength is needed with low weight; for example, helicopter rotor blades, airplanes and kayaks.

Quick Test

1. Why were gold and silver amongst the first elements discovered?
2. How would you create small crystals in ceramic plates?
3. What is a composite?
4. Explain how to make a polymer.

Key Words

ceramic
monomer
polymer
composite

Using our Earth Sustainably

You must be able to:

- Describe the structure of the Earth
- Explain rock formation and the rock cycle
- Describe the composition of the Earth and its atmosphere.

The Structure of the Earth

- The Earth is a planet in orbit around the Sun.
- All of the chemicals used in industry come from the Earth.
- Rocks make up the solid crust of the Earth.
- The rocks in the crust contain chemical compounds and elements that can be extracted and used.

The Rock Cycle

- There are three types of rocks, classified according to how they formed.
- **Igneous** rock is formed from liquid rock.
  - If the liquid rock cools rapidly, the igneous rock formed will have small crystals, e.g. basalt.
  - If the liquid rock cools slowly then the igneous rock will have large crystals, e.g. granite.
- When rocks have been subjected to weathering and erosion, pieces of rock break off.
- The sediments formed eventually settle and are subjected to large pressures from the rock above.
- When material deposited in this way forms rock it is called **sedimentary** rock.
- Any material that can be deposited can lead to sedimentary rock, e.g. eroded rock material, or the calcite skeletons of microorganisms that lived in the sea.
- Due to the movement of the tectonic plates, rock that was on or near the surface can be moved closer to the Earth's core.
- Subjected to intense heat and pressure, the structure of the rock alters.
- The altered rock becomes **metamorphic** rock.

Structure of the Earth

core (consists of iron and nickel)

crust

mantle

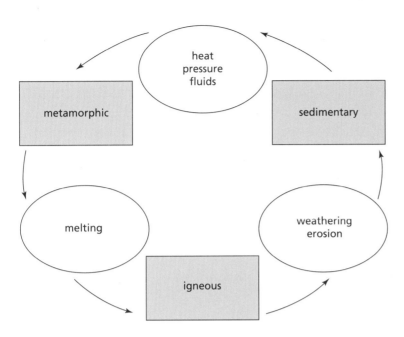

Sedimentary rock

Igneous rock

Metamorphic rock

Composition of the Earth

- The chemicals in the Earth's crust tend to be insoluble compounds. If they were soluble, the ground would dissolve whenever it rained!

Element	Abundance in the Earth's crust (%)
Oxygen	46
Silicon	28
Aluminium	8
Iron	5
Other elements	13

- The atmosphere of the Earth contains very light compounds that exist as gases, as well as very small solid particles that are light enough to float in the air.

Element	Abundance in the atmosphere (%)
Nitrogen	78
Oxygen	21
Argon	0.9
Carbon dioxide	0.04
Other trace gases	0.06

Key Point

The composition of elements is very different in different parts of the Earth. It is important to specify which part you are describing.

Quick Test

1. What are the three types of rock?
2. Why do chemicals in the Earth's outer layer tend to be insoluble?
3. Draw the rock cycle.

Key Words

igneous
sedimentary
metamorphic

Using our Earth Sustainably

You must be able to:

- Describe the carbon cycle
- Explain the impact of human activity on the atmosphere
- Describe what is present in the atmosphere and how it has changed over time
- Suggest why the Earth is a source of limited resources.

Quick Recall Quiz

The Carbon Cycle

- The element carbon (C) is common on Earth.
- Carbon is reactive and can form up to four bonds with different elements, often forming chains.
- The different reactions that carbon takes part in mean that carbon atoms move through the **carbon cycle**.
- Changes to parts of the cycle will have an impact on other parts of the cycle.

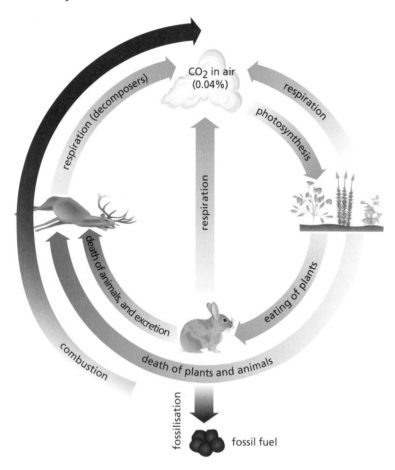

Human Activity and Climate Change

- Human activity can alter the balance in the carbon cycle.
- By removing and burning **fossil fuels** (coal, oil and natural gas), carbon that was trapped for millions of years is now released into the atmosphere as carbon dioxide.

> **Key Point**
>
> Levels of CO_2 are globally the highest they have been for the past 3 million years.

- This carbon dioxide acts as a greenhouse gas, trapping heat from the Sun.
- The more carbon dioxide released, the hotter the planet becomes – the consequence of this is climate change.
- In different parts of the world, the oceans get warmer.
- This causes changes to currents and wind patterns, and unpredictable and extreme weather patterns can occur.
- Humans produce carbon dioxide through a number of activities:
  - factories or transport (e.g. planes, cars) using fossil fuels
  - cutting down and burning forests.
- Scientific **consensus** indicates that the amount of carbon dioxide in the atmosphere needs to be reduced.
- Before industrialisation in the 1800s the amount of carbon dioxide in the atmosphere was 0.028%.
- In 2013 the level reached 0.04% for the first time in the past 3 million years.

Limited Resources

- Although the planet looks enormous to us and seems to have an endless supply of resources, this is not true:
  - Many elements that we find most useful are also rare
  - Many of the components used in mobile phones and tablets are made using rare earth metals
  - The majority of the energy we use comes from non-renewable sources, such as coal, oil and natural gas
  - Most items in households of the Western world are made from the products of crude oil.
- The finite reserves of oil are being used up, and since oil takes millions of years to form, we cannot make more. This means that we need to recycle items.
- **Recycling** involves extracting parts of a used product and making them available for other processes or products, e.g. recycling paper involves collecting the used paper, sorting it and then treating it to make recycled paper.
- Of course, recycling uses energy to extract the materials.
- Manufacturers are increasingly being required by law to make the extraction and recycling of materials easy.

> **Key Point**
>
> Although the media often show both sides of an argument, it is important to recognise when one side has far more scientific evidence than another.

> **Key Point**
>
> Sustainable development helps us to make responsible decisions by ensuring projects improve people's standard of living by having nice things today, but also protecting the environment from damage from pollution and climate change in the future.

> **Quick Test**
>
> 1. What do scientists believe is causing CO_2 levels to increase?
> 2. What is recycling?
> 3. Draw the carbon cycle.
> 4. How are humans affecting the Earth's atmosphere?

> **Key Words**
>
> carbon cycle
> fossil fuels
> consensus
> recycling

Explaining Physical Changes

1 The diagram below shows the arrangement of atoms in four different substances.

A

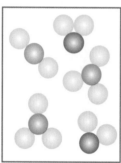

B

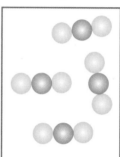

C

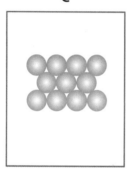

D

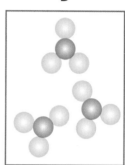

a) Which box has a substance that is a mixture of different compounds? [1]

b) Which box has a substance that could be carbon dioxide? [1]

c) Which box has a substance that is a pure element? [1]

2 It is winter and Sylvia has gone to look at her fishpond. The water pump is switched off. There is a layer of ice over the top of the pond. Despite the ice, the fish in the pond are alive and are able to swim.

a) Explain what property, unique to water, has enabled the fish in the pond to survive. [3]

b) Sylvia's sister Hilary suggests breaking the ice and switching the water pump back on.

Explain why this could lead to the death of the fish in the pond. [3]

Explaining Chemical Changes

1 **a)** Write the word equation for the reaction of sodium and chlorine. [2]

    b) Write the chemical equation for the reaction of sodium, Na, and chlorine, Cl_2. Include state symbols. [2]

2 Hydrogen peroxide is a chemical that acts as a bleach.

Hydrogen peroxide decomposes to form water and oxygen. The reaction happens more quickly when manganese dioxide is present.

    a) Write the word equation for the decomposition of hydrogen peroxide. [2]

    b) Manganese dioxide is neither a reactant nor a product. What is the chemical name for this type of chemical? [1]

3 In 2012 Feliz Baumgartner floated in a helium balloon until he reached a height of 39 km before jumping out.

    a) The diagram below shows a balloon that is floating.

Copy the diagram and draw arrows inside the balloon to represent the pressure of the gas inside and arrows representing the air pressure outside the balloon. [2]

    b) Draw the arrangement of particles inside the balloon. [1]

Obtaining Useful Materials

1 The list below is a shortened reactivity series. Use it to answer the following questions.

potassium
sodium
calcium
aluminium
carbon
iron
tin

a) Which of the metals can be extracted from their ores by reacting with carbon? [2]

b) What would be the products of the displacement reaction between carbon and iron oxide?

Complete the word equation.

carbon + iron oxide ⟶ _____ + _____

[2]

2 Name the small units that join together to form a polymer. [1]

3 The volcanic rock basalt is one of the most common rocks on the Earth. It has small crystals that can only be seen clearly using a microscope.

Suggest how quickly basalt cools. [1]

4 Colin is studying the architecture of the Roman Civilisation. He reads that the Romans were the first to make concrete. Concrete is made from stones and cement.

Explain why concrete is a composite material and suggest why the Romans chose to use concrete in their buildings, rather than cement or stones on their own. [2]

Using our Earth Sustainably

1   The Earth consists of layers. Copy the picture below and add labels to identify each layer.   **[3]**

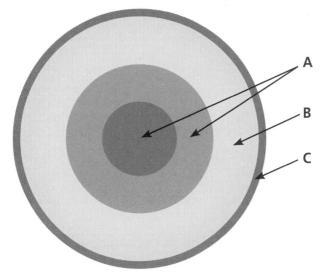

2   The rock cycle below is missing some labels. Copy the diagram and fill in the gaps.   **[5]**

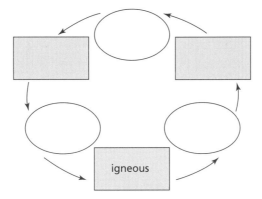

3   Scientists have provided evidence that man's activities are a direct cause of climate change.

   a)   What is climate change?   **[1]**

   b)   Which of the following are direct causes of climate change?

     i)   using nuclear power stations

     ii)   burning coal in power stations

     iii) geothermal energy

     iv) petrol powered cars   **[2]**

4   Explain why it is important to recycle resources.   **[2]**

Forces and their Effects

You must be able to:

- Describe some causes and effects of forces
- Explain how objects can be affected by forces
- Use force arrows to show forces acting on objects
- Explain the concept of moments
- Describe and explain Hooke's Law.

Quick Recall Quiz

What are Forces?

- **Forces** are pushes or pulls between two objects.
- Force arrows can be drawn to show the direction that the force is acting in:

- When forces act on an object, there will be a consequence:
  - The object may become deformed, e.g. stretched or squashed
  - The object may warm up due to rubbing and the friction between the surfaces
  - The object may be pushed out of the way
  - The object may provide resistance to the motion of water or air.

Balanced and Unbalanced Forces

- Forces act in opposite pairs and force arrows are drawn to show this.
- The forces acting on an object can either be:

Balanced	Unbalanced
This means that there will be no change in the object.	This means that there will be a change in the object.
The force of gravity pulling down on the books is balanced by the reaction force applied upwards by the shelf.	The driving force of the cycle is greater than the air resistance and friction – the force is unbalanced. The cyclist is accelerating.

> **Key Point**
>
> You should clearly identify the direction forces act in using arrows, and the size of the force by the length of the arrow.

Measuring Forces

Forces are measured in **Newtons** (N).
The stronger the force, the greater the value in Newtons.
Forces can be measured using a Newton meter.
The size of a force can also be measured by the size of the force's effect.
When an object is stretched or squashed, the change in its length can be measured.
The greater the force applied, the greater the stretch or squashing force.
There is a linear relationship between the size of the force and the resulting stretch (i.e. they change in proportion to each other) as long as the elastic limit is not exceeded.
This means that for every unit of force, there will be the same effect on the stretch (e.g. if 1 N causes 3 cm stretch, then 2 N causes 6 cm stretch, 10 N causes 30 cm, etc.) as long as the elastic limit is not exceeded.

Pivots and Moments

A **pivot** is the turning point, or hinge point, of a simple machine.
A **moment** is the turning effect of a force, and has the unit Newtonmetre (Nm).
moment = force × distance from pivot.
The numbers in the diagram should be used to calculate the moment.

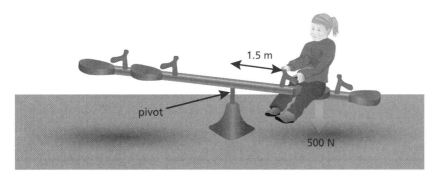

1.5 m

pivot

500 N

The further away from the pivot, the greater the moment.
This is why a long-handled screwdriver can be used to open a tin of paint.

Hooke's Law

Hooke's Law states that the stretch of a spring will be directly proportional to the force applied (i.e. there is a linear relationship between them).

Graph showing Hooke's Law

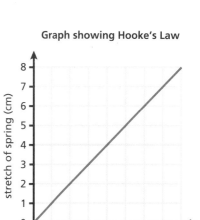

stretch of spring (cm)

force (N)

Quick Test

1. How do we describe a pair of forces if one force is larger than the other?
2. What is the unit for force?
3. What do force arrows tell us?
4. What is meant by the term 'moment'?

Forces and their Effects

Physics

You must be able to:

- Explain simple effects of forces on speed and direction
- Calculate speed from distance and time
- Explain how work is done when a force acts on an object and changes it
- Explain how deformation involves work.

Forces and Motion

- An object will move in the direction a force is applied.
- The larger the force, the faster the object will start to move.
- Applying a large enough opposing force to an object's movement will cause the object to slow down and stop, for example, a football net catching a football.

- To change an object's **speed**, a force has to be applied.
- For an object to speed up, more force is needed in the direction of movement.

- For an object to slow down, the greater force has to be in the opposite direction to the direction of movement.

> **Key Point**
>
> When forces are balanced an object will continue at the same speed. An object that isn't moving has a speed of 0 m/s. If it isn't already moving, it will continue to not move.

Car travelling on tarmac at constant speed 70 mph. The forces are balanced.

There is greater friction between the gravel and the car tyres.
To travel at 70 mph, more force has to be produced by the engine.

Speed, Distance and Time

- Speed is a measure of the distance that an object travels in a given time.
- speed = distance ÷ time
- The faster the speed, the further the object travels in a set period of time.
- If an athlete ran 100 m in 10 s then:
    speed = 100 m ÷ 10 s
          = 10 m/s
- The magnitude of the change in speed of an object depends on the size and direction of the force that is applied.

Work and Energy Changes

- In science, work is done if a force has acted on an object and there has been a change in that object.
- So, if an object did not move at all, no work will have been done.
- But, if the object moved faster or slowed down, then work will have been done.
- If the object becomes deformed (squashed or stretched) then work will also have been done.
- Work is measured in Newton metres (Nm), or Joules (J).

Key Point

Work is only done if the object changes speed or shape.

Quick Test

1. Give a metric unit for speed.
2. What is meant by the term 'work'?
3. A car drives 48 km in 40 minutes. What is its speed?

Key Words

speed
work
Joule

Physics

Exploring Contact and Non-Contact Forces

You must be able to:

- Explain non-contact forces between magnets and static electricity
- Explain electrostatic attraction and repulsion
- Describe the effects of gravity across space.

Non-Contact Forces

- A contact force is applied by one object touching another.
- Forces can also act over a distance. These are called non-contact forces, e.g. magnetism, static electricity and gravity.

Magnetism

- Magnets have two poles, North and South.
- The opposite poles of a magnet **attract**.
- The same poles of a magnet **repel**.

> **Key Point**
>
> Always remember to write both poles on diagrams of magnets.

Magnets attracting and repelling

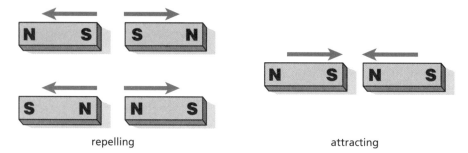

repelling   attracting

Static Electricity

- Static electricity occurs when an object gains or loses electrons (electrons have a negative charge).
- If the object gains electrons it becomes negative; if it loses electrons it becomes positive.
- If two objects with the same charge are brought together, they will repel each other.
- If an object with a negative charge is brought near an object with positive charge, they will be attracted.

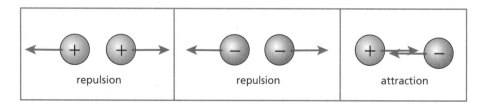

repulsion   repulsion   attraction

Electric Fields

- When an electric current passes through a wire it produces an **electric field**.
- Electric fields act across the spaces between objects that are not in contact with one another.

Charged parallel plates

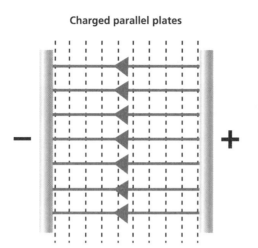

The diagram shows two parallel plates, one of which a has positive charge and the other a negative charge. Any charged object put in the space between them will experience a force. For example, a positively charged object will be repelled by the positive plate and attracted by the negative plate.

Gravity

- Gravity is a force exerted by one object on another when they are near each other.
- On the Earth everything is pulled to the Earth's centre.
- The Earth has a gravitational field strength of 10 N per kg. This means every kilogram on Earth has a force of 10 N acting on it.
- Every object with mass has a **weight**, measured in N, if it is in a gravitational field.
- weight = mass × gravitational field strength
- Mass is the measure of all the matter in an object, and has the units kg.
- On other planets and moons the gravitational field strength will be different. The weight of an object will therefore be different on each planet, but the mass will stay the same.
- All objects with mass have a gravitational pull, even people. As the Earth is so massive, we do not notice these gravitational pulls between objects.
- The further away you are from the centre of mass causing the gravitational field, the weaker the gravitational force. For example, a person standing on the top of Mount Everest will experience slightly less gravitational pull than a person standing in Trafalgar Square, London.

> **Key Point**
>
> The International Space Station orbits Earth at around 400km above the surface. Although objects on board appear to be weightless, they are not; everything is falling at the same speed.

> **Key Point**
>
> Gravitational field strength is measured in N/kg.

Quick Test

1. What is the difference between weight and mass?
2. Why does weight decrease with distance from the Earth?
3. What forms around a wire when electric current flows through it?
4. Draw two magnets repelling and attracting.

> **Key Words**
>
> attract
> repel
> electric field
> weight

Exploring Contact and Non-Contact Forces

You must be able to:

- Explain how pressure acts in the atmosphere, in liquids and in solids
- Explain pressure as the effect of force over area
- Explain why objects float and sink.

Pressure

- Forces can act over an area, in all directions.
- pressure = force ÷ area
- Pressure is measured in Newtons per metre squared (N/m^2).
- A person wearing snowshoes exerts less pressure on the ground than a person of the same mass wearing ice skates.

> **Key Point**
>
> Make sure that the unit is written N/m^2 and not Nm, which is the unit for the moment of a force.

Atmospheric Pressure

- The atmosphere on the Earth is exerting a pressure on all objects on the surface.
- As an object gets higher, atmospheric pressure reduces, because there is less atmosphere above it pushing downwards.

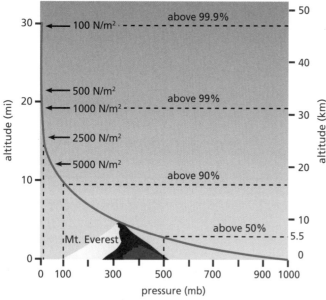

pressure at surface = 101,000 N/m^2

Pressure in Liquids

- With liquids, pressure increases with depth.
- The deeper an object gets, the greater the force acting on it due to the weight of the liquid above.
- The pressure at the top of the bottle is less than the pressure towards the bottom. Water pours out of the bottom hole much faster than the hole at the top.

Water coming out of 3 holes in a bottle

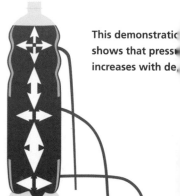

This demonstratic shows that presse increases with de,

Upthrust

- When an object is placed into water, the water exerts a force in the opposite direction to the weight of the object.
- The term for this is **upthrust**.
- The more dissolved salts in the water, the greater the upthrust.
- Many ships have a series of lines, called **plimsoll lines**, painted on the hull to show how deep the ship will sink in different waters. In fresh water a ship will sink lower whilst in very salty water, such as in the Dead Sea, the ship will sink less due to the increased upthrust.

Floating and Sinking

- An object will float if the upthrust equals the weight of the object.
- If the weight is greater than the upthrust, the object will sink.
- Another important consideration is the **density** of the object. The greater the density, the more likely it will be that the weight will overcome the upthrust from the liquid and the object will sink.
- Ships float even though they may weigh thousands of Newtons because the weight is spread over a large area.

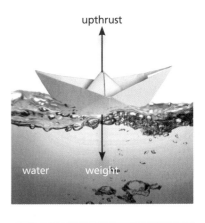

upthrust

water   weight

Key Point

Remember – when the forces acting on an object are balanced it will continue to do what it was doing. If it is floating, it will continue to float.

Upthrust / downthrust of blocks in water

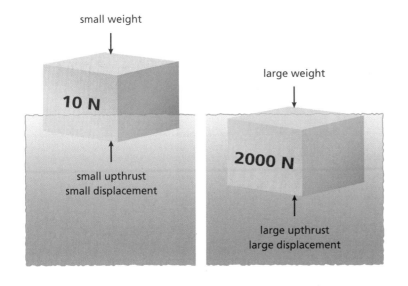

small weight

10 N

small upthrust
small displacement

large weight

2000 N

large upthrust
large displacement

These two blocks are the same size but the one on the right is denser so it is floating lower in the water. If it was denser than water, the upthrust would not be enough to support it and it would sink.

Quick Test

1. Why does atmospheric pressure decrease with height?
2. What is the unit for pressure?
3. If an object is floating with weight of 1000 N, what is the upthrust?
4. For an object to float higher in water, what must be increased?

Key Words

pressure
upthrust
plimsoll line
density

Obtaining Useful Materials

1 Which of the following metals was discovered thousands of years ago? [1]

Gold

Aluminium

Iron

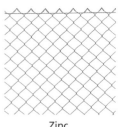
Zinc

2 **a)** Predict which **two** of the following reactions will take place.

i) copper + aluminium nitrate   **ii)** iron + copper sulfate

iii) iron + potassium nitrate   **iv)** copper + silver nitrate [2]

b) For each of the reactions you selected, write the complete word equation for the reaction. [2]

3 The photo shows a blast furnace. A blast furnace is used to extract iron from its ore.

Explain what happens when carbon and iron oxide are reacted together in the blast furnace. [3]

Using our Earth Sustainably

1 Match the rock description to the type of rock:

Rock type	Rock Description
Igneous	may contain fossils
Sedimentary	often have a striped appearance and can be used to make statues
Metamorphic	is very hard and made of lots of small crystals

[3]

2 Copy the table and use the words below to complete the gaps, showing the abundance of elements in the Earth's crust.

aluminium   iron   oxygen   silicon

Element	Abundance (%)
	46
	28
	8.0
	5.0
Other elements	13

[4]

3 A group of friends are talking about recycling.

Kalum: I think that recycling takes up energy so is not worth doing.

Ruth: Recycling leads to rare materials being extracted and used for other purposes.

Carlos: There are plenty of resources. This is just an excuse to raise taxes.

Arwen: We are running out of oil which means we will no longer be able to make plastics.

Kalum   Ruth   Carlos   Arwen

a) Which two friends are giving reasons **for** recyling? [1]

b) Write a counter argument to address Kalum's point. [2]

Forces and their Effects

1 Look at the Newton meter.

a) Select the units that force is measured in. [1]

    grams   **kilograms**

    Newtons   **Newtons per kilogram**

b) State the reading on the meter. [1]

c) Explain what is causing a force to be applied to the meter. [1]

2 Sandy is playing with balancing scales. On the left-hand scale she has a ball and on the right she adds mass until it balances.

a) If it balances with 200g on the right, what is the mass of the ball in grams? [1]

b) She now removes the ball and puts two identical blocks on the left-hand scale and 300 g on the right hand scale to balance it. What is the mass of one block in grams? [1]

3 Arran is training to run the 800 m in the Olympics. In her last race she won in 1 min 48 s. What was her average speed for the race? [2]

4 Joss is prising open the lid on a tin of paint. Calculate the **turning moment** for the force applied to the screwdriver.

Show your working and give the correct unit. [3]

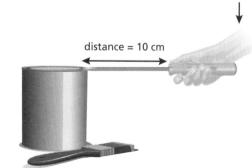

force = 15 N

distance = 10 cm

Exploring Contact and Non-Contact Forces

1 Fiesa brings two magnets together. They repel.

a) Copy and complete the diagram to show the poles on the magnets.

Magnet 1   Magnet 2

| N | S |

| | |

[2]

b) What would Fiesa have to do to make the magnets attract?   [2]

2 Taking the gravitational field strength on Earth to be 10 N/kg, what would each of the following masses weigh?

10 kg   15.5 kg   2000 g   [3]

3 The bottle of water is full. The top is open and the bottle has three holes made at points A, B and C.

a) Draw what would happen to the flow of water at points A, B and C on the bottle.   [1]

b) Explain why the flow at points A, B and C is different.   [2]

4 Amy and George are both the same weight, 600 N. Amy wears snowshoes with an area of 0.025 m² and George wears skis with an area of 0.035 m². Calculate the pressure exerted by each of them on the snow.   [2]

Physics

Motion on Earth and in Space

You must be able to:

- Interpret distance–time graphs
- Explain and apply concepts of balanced forces and equilibria to analyse stationary objects
- Explain relative motion.

Describing Motion

- The motion of an object (the journey it takes) can be described by drawing a distance–time graph.
- The axes of the graph must be labelled correctly.
- Time is plotted on the x-axis.
- Distance is plotted on the y-axis.

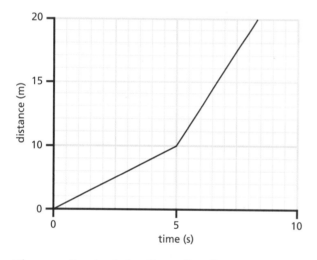

- The gradient of the line of a distance–time graph represents the speed the object is travelling at:
  - A steeper line means more distance is covered in the same time, i.e the speed is greater
  - A shallower line means less distance is covered in the same time, i.e. the speed is lower
  - When the line is horizontal it means the object is not moving at all. It has stopped.
- A more complex distance–time graph appears below:

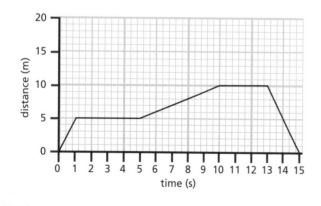

Relative Motion

- The motion of objects is always **relative** to the **observer**.
- If two trains, A and B, are travelling in the same direction and at the same speed on tracks parallel to one another it would appear to an observer on either train that both trains were at a standstill.
- If the trains were travelling on parallel tracks towards each other at the same speed, then an observer on either of the trains would get the impression that the other train was travelling at twice the speed of their train.
- To calculate relative velocity to an observer:
  - If the object is moving towards the observer, add the speeds.
  - If the object is moving away from the observer, subtract the speeds.
- So for trains moving parallel to each other:

    Observer ➡ 10 m/s
    Object ➡ 10 m/s

    relative velocity = 10 m/s – 10 m/s = 0 m/s
- For trains moving in opposite directions:

    Observer ➡10 m/s
    Object ⬅ 10 m/s

    relative velocity = 10 m/s + 10 m/s = 20 m/s

> **Key Point**
>
> The position of an observer may influence what they see. Everything is relative!

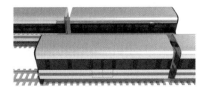

> **Key Point**
>
> Velocity is the same as speed, but with direction added.

Forces in Equilibrium

- When forces act in opposite directions to each other and are the same size, they are balanced or in **equilibrium**.
- When the forces are balanced, the object will continue to do what it is doing:
  - If it is moving, it will move at a constant speed.
  - If it is stationary, it will stay stationary.
- A spring with a 10 N weight attached will stretch until the force of the spring pulling the weight upwards (the **reaction force**) equals 10 N.

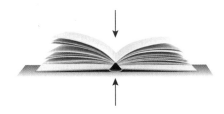

> **Quick Test**
>
> 1. If an apple produces 1 N of force on a desk, it is not moving. What is the reaction force?
> 2. What can a distance–time graph tell us?
> 3. One car travels left at 70 mph and another travels right at 60 mph. What is the relative velocity?
> 4. What does the slope tell you on a distance–time graph?

> **Key Words**
>
> relative
> observer
> equilibrium
> reaction force

Motion on Earth and in Space

Physics

You must be able to:

- Explain differences in gravity on different planets
- Explain why the Earth has seasons
- Explain why day length varies on both the Earth's hemispheres.

Space and Gravity

- Gravity is a force that gives objects weight.
- Every object with mass has gravity.
- Gravity, however, is a relatively weak force and a very large mass is needed before the gravitational pull is noticed.
- Therefore, all planets and moons (not just Earth) have a gravitational pull.
- Weight is calculated by the formula:

weight = mass × gravitational field strength (g)

- The Earth has a **gravitational field strength** (g) of 10 Newtons per kilogram, which is written as: **g = 10 N/kg**
- On other planets and moons the value of g will differ.

> **Key Point**
>
> Do not confuse mass and weight. Weight is a force and changes depending where you are. Mass always stays the same.

Planets, showing approximate gravitational field strength values (N/kg)

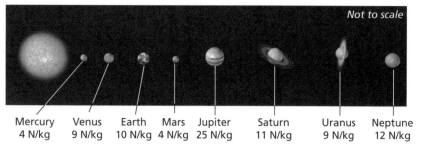

Not to scale

Mercury	Venus	Earth	Mars	Jupiter	Saturn	Uranus	Neptune
4 N/kg	9 N/kg	10 N/kg	4 N/kg	25 N/kg	11 N/kg	9 N/kg	12 N/kg

- The gravitational field strength of the Earth keeps the Moon in orbit, whilst the much larger gravitational field strength of the Sun keeps the Earth and all the other planets in orbit around it.
- The Sun is our closest star and it orbits the centre of our galaxy, the Milky Way.
- The Milky Way is filled with billions of other stars, all orbiting the galactic centre due to the gravitational pull from a black hole in the centre.
- Our galaxy is only one of billions, all exerting a gravitational pull on each other.
- Distances in space are enormous, so scientists use the astronomical unit for distance, the **light year**.
- A light year is the distance that light travels in 1 earth year, 9,460,730,472,580,800 m or 9.5 trillion km.

The Earth in Space

- The Earth spins on its axis once every 24 hours.
- The Earth is tilted on its axis, so at different times of the year the Northern and Southern hemispheres receive different amounts of radiation from the Sun. This leads to seasons.
- Day length varies in the two hemispheres as the planet orbits the Sun.

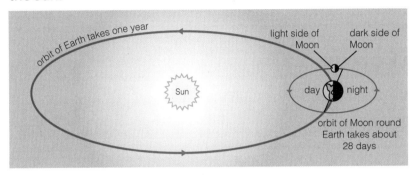

- An **equinox** is when both hemispheres of the Earth receive the same amount of light, so day and night are the same length wherever you are on the planet.
- There are two equinoxes a year, on the 21st March and 21st September.
- Between 21st March and 21st September, days are longer than nights in the Northern Hemisphere. In the Southern Hemisphere the opposite is true.
- After 21st September, day length shortens in the Northern Hemisphere and lengthens in the Southern Hemisphere.
- For the Northern Hemisphere 21st June is the longest day (summer **solstice**) and 21st December the shortest day (winter solstice). This is reversed for those in the Southern Hemisphere.

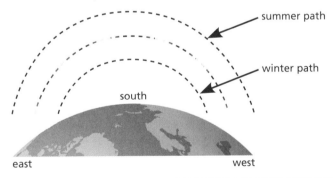

Key Point

If the Earth was not tilted then there would be no seasons. There would be no difference in day length across the planet.

Key Point

The dates given for the equinoxes and solstices are the average. The precise dates vary according to the year.

Quick Test

1. How long does daylight last on 21st March (the spring equinox) in the Northern Hemisphere?
2. Why do we get seasons?
3. If a person weighed 800 N on Earth, what would they weigh on Jupiter, where the gravitational field strength is 25 N/kg?
4. Suggest why *g* is only 4 N/kg on Mercury.

Key Words

gravitational field strength
light year
equinox
solstice

Energy Transfers and Sound

You must be able to:

- Understand and apply the model of energy transfer to various contexts
- Explain the role and significance of fuels.

Energy Changes Due to Forces

- When force is applied, work done can be calculated using the formula:

> **work done = force × distance**

- Increasing the distance to the pivot means less force is applied, but over a larger distance. Reducing the distance increases the force used. The amount of work always remains the same.
- Similarly, gears increase the distance to the pivot, with a smaller gear turning many times, making a larger gear move.
- When an object changes its motion, other changes are experienced, e.g. dropping an object causes changes in the energy in the object. The energy is transformed from **gravitational potential energy** to **kinetic energy**, e.g. dropping an object causes energy to be transferred from one store to another. The energy is transferred from the gravitational potential store to the kinetic energy store.

Cogs turning

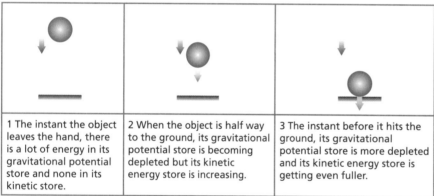

1 The instant the object leaves the hand, there is a lot of energy in its gravitational potential store and none in its kinetic store.	2 When the object is half way to the ground, its gravitational potential store is becoming depleted but its kinetic energy store is increasing.	3 The instant before it hits the ground, its gravitational potential store is more depleted and its kinetic energy store is getting even fuller.

- A dynamo is a device that transfers energy from the kinetic store of whatever is driving it, by means of electricity, to another store. If it is being used to charge a battery, the store is the chemical potential store in the battery; if it is being used to power a light bulb, energy is being transferred to the environment by light and thermal energy.
- The more efficient a transfer, the more energy is transferred in a way that is useful.

> **Key Point**
>
> Often some of the energy is 'lost' as heat and sound. We say it is transferred.

Energy Changes Due to Altering Matter

- When forces are applied to an object they will change its shape.
- When **compression** is applied to a spring, energy is transferred to its elastic potential energy store.

- When the spring is released, it returns to its original shape and energy is transferred out of its elastic potential store.
- The elastic potential energy is transferred to the spring's kinetic energy store.
- Fuel has a **chemical potential energy** store.
- When the fuel is ignited, energy is transferred to the thermal store of the surroundings:

fuel + oxygen ⟶ carbon dioxide + water + energy

- There are many types of fuels, storing different amounts of chemical potential energy.
- Most of the fuels in current use derive from crude oil although the number of alternatives, such as biodiesel, are increasing.
- The metabolism of food is similar to the combustion of fuel.
- The process inside cells is called respiration. It happens more slowly than combustion and involves food (the fuel) reacting with oxygen:

food + oxygen ⟶ carbon dioxide + water + energy

- When hot and cold objects come together, thermal energy will transfer into the cooler object, until both objects are the same temperature.

> **Key Point**
>
> You can't let the cold in, only the heat out!

Energy Changes Due to Vibrations and Waves

- The Sun warms the Earth by radiation.
- Radiation is the transfer of thermal energy via waves (infrared) that can travel in a vacuum.
- The objects do not need to be touching to receive the thermal energy.
- **Vibration** of atoms occurs when objects get hot.
- The more kinetic energy, the more vibrations there are and the hotter they get. As energy is transferred to a material, the more the atoms vibrate. Their kinetic energy store increases and the temperature of the material increases.

Energy Changes Due to Electricity

- In an electrical circuit the battery has a chemical potential store.
- When the circuit is completed, energy is transferred from the chemical potential store in the battery via electricity. The store it is transferred to depends on the other components in the circuit. If there are bulbs in the circuit, they will transfer energy to the environment by light and thermal energy.

> **Quick Test**
>
> 1. What is the word equation for the combustion of fuel in oxygen?
> 2. Describe the differences in reaction rate between combustion and respiration.
> 3. What is meant by radiation in terms of energy transfers?
> 4. An object has gravitational potential energy of 1000 J. What will its maximum kinetic energy be when dropped?

> **Key Words**
>
> gravitational potential energy
> kinetic energy
> compression
> chemical potential energy
> vibration

Energy Transfers and Sound

Quick Recall Quiz

You must be able to:

- Explain how sound waves behave
- Explain how different animals hear
- Explain how sound is produced.

Sound Waves

- Sound waves carry energy through a medium.
- The medium must have particles to transfer energy as sound.
- The closer the particles, the faster the energy can be transferred.
- Sound travels fastest in solids, slower in liquids and slowest in gases.
- The speed of sound in air is 340 m/s, in water it is 1500 m/s and in solids such as wood, 4000 m/s (approximately).
- The **frequency** of a wave is measured in hertz (1 Hz = 1 wave per second).
- Musical notes have separate frequencies.
- Animals can hear a range of sounds.
- As humans get older, the range of hearing decreases, with the highest frequency sounds being lost first.
- Objects produce sound when they vibrate. For example:
  1. A guitar string when plucked causes a vibration in the wire.
  2. The wire hits air particles and causes them to move.
  3. The air particles collide with other air particles. Eventually they make the ear drum vibrate at the same frequency.
- Loudspeakers work in a similar way:
  1. Music is converted to an electrical signal which causes electromagnets to move a fabric skin or membrane.
  2. The fabric skin's motion causes air particles to move.
  3. Eventually the sound waves reach the ear.
- Microphones work in the opposite way:
  - Sound waves produced by an instrument or voice hit a membrane attached to magnets
  - The motion of the magnets causes a changing electrical signal that can be recorded and used to make a loudspeaker move.

> **Key Point**
>
> A vacuum does not contain particles, so sound cannot travel in a vacuum.

Hearing range of animals

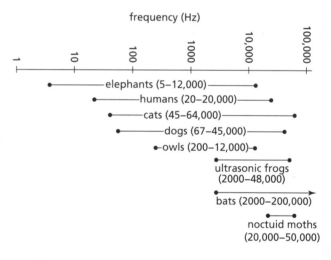

Sound from loudspeaker

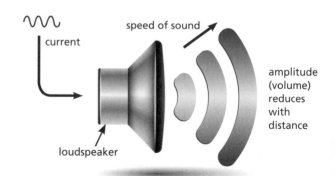

Using Sound

- Very high frequency sound (greater than 20000 Hz) can be used to clean objects, e.g. jewellers can clean jewellery using an ultrasonic bath. The high frequency vibrations cause minute particles of dirt to be displaced.
- **Ultrasound** is also used in medicine, e.g. to treat injuries and to detect the movement of a developing foetus in the mother's uterus.

Echoes and Absorption

- Sound waves behave like any other wave, which means that they can be reflected:

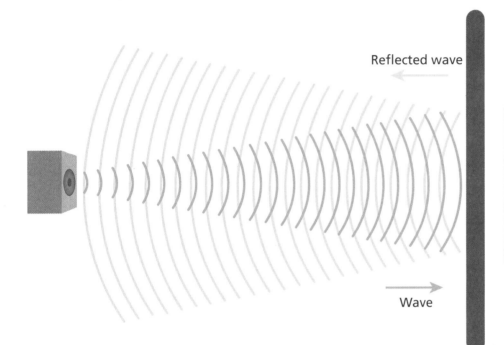

Reflected wave

Wave

- When sound waves reflect straight back an **echo** is heard.
- When echoes would be a problem, e.g. in a music recording studio, steps are taken to prevent any echoes from spoiling the recording.
- Putting special angled tiles onto the walls and ceilings ensures that the sound waves keep reflecting within the tile until the energy has been **absorbed**.

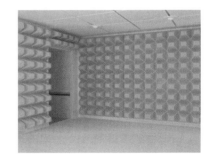

Quick Test

1. What does it mean if a wave has a frequency of 50 Hz?
2. Explain how echoes occur.
3. Suggest how to reduce echoes in a sound studio.
4. Explain how a loudspeaker works.

Key Words

frequency
ultrasound
echo
absorb

Forces and their Effects

1 An astronaut is stationary in space and the following forces are applied. For (a), (b), and (c), decide whether the astronaut will move or not. [3]

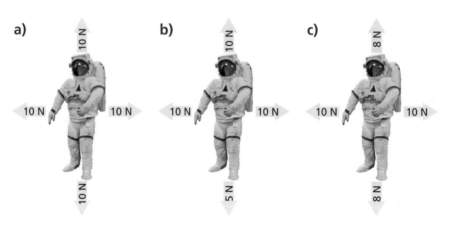

a)

b)

c)

2 Ges is investigating whether a spring obeys Hooke's Law. He hangs masses from a spring and measures the length the spring stretches each time. The table below shows the results for Ges's experiment.

Mass (g)	Weight (N)	Stretch (cm)
0	0	0
100	1	2
200	2	4
300	3	6
400	4	8
500	5	10
600	6	
700	7	
800	8	16

Ges did not record a result for 600 g and 700 g. Predict what the stretch of the spring should have been (in cm) for these. [2]

3 The International Space Station (ISS) orbits the Earth at an average speed of 27,600 km/h.

Taking the distance of one orbit of the Earth as 42,927 km, how long does it take the ISS to orbit the Earth once? Show your working out. [3]

Exploring Contact and Non-Contact Forces

1 Which of the following forces are examples of non-contact forces?

a) two teams of people having a tug of war   **b)** a person unscrewing a bottle of fizzy drink

c) gravity pulling a seed towards the ground   **d)** static charge attracting dust   [2]

2 Charlotte rubs a polythene rod with a cloth and holds it close to a stream of flowing water from the tap.

Normal flow
of water

Flow changes with
a charged rod

a) Suggest why the water is attracted to the polythene rod.   [2]

b) Charlotte hangs the charged rod by a piece of string so that it is suspended horizontally and is free to swing. She then brings another rod with the same charge close to it. What will happen to the hanging rod?   [1]

3 This question is about a bicycle.

a) Identify two places on the bicycle where friction should be as great as possible.   [2]

b) Identify two places on the bicycle where friction should be as low as possible.   [2]

Motion on Earth and in Space

1 Lisa is walking her dog. The distance–time graph below shows Lisa's journey.

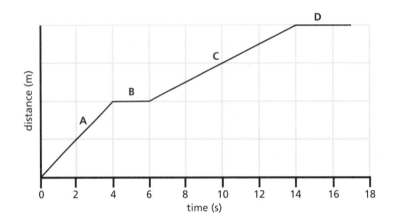

Each of the following describes part of Lisa's journey. Choose which of the labels A–D match with the parts of Lisa's journey given below:

Lisa stopped for 2 min   **Lisa walked for 8 min**

Lisa jogged for 4 min   **Lisa stopped for 3 min**   [4]

2 Two cars are involved in a head-on collision. Luckily the occupants of both cars were not injured.

If both cars were travelling at 40 km/h, what would the relative velocity have been?   [1]

3 The closest star to Earth outside the solar system is Proxima Centauri, which is 4.2 light years from the Earth.

a) What is a light year?   [1]

b) Taking the speed of light to be 300,000 km/s in a vacuum, calculate how far away Proxima Centauri is from the Earth in km. Show your working.   [3]

4 The diagram below shows the path the Sun takes in the Northern Hemisphere in June.

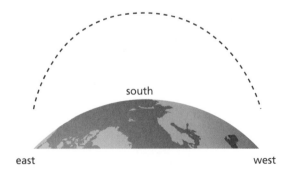

Copy the diagram and add the path that the Sun will take in December.   [2]

Energy Transfers and Sound

1 A bungee jumper jumps off a tower.

a) At what point does the bungee jumper have the most kinetic energy? [1]

b) At what point does the bungee jumper have the most gravitational potential energy? [1]

2 The equation below is the word equation for the combustion of a fuel burned in excess oxygen.

> **Fuel + A ⟶ B + C +** energy

a) What are A, B and C? [2]

b) What is the name of the process in cells that is equivalent to combustion? [1]

3 Which of the following is the way that the Sun heats the Earth?

a) conduction

b) convection

c) radiation

d) nuclear [1]

4 Jamie is talking to his wife Linda. He is standing in the kitchen and Linda is in the next room.

Explain why Jamie and Linda can hear each other talking, even though they are not in the same room. [2]

Magnetism and Electricity

Quick Recall Qui

You must be able to:

- Explain electric current as a flow of charge
- Apply concepts of potential difference and resistance
- Describe and measure current in series and parallel circuits
- Use formulae to calculate resistance, current, voltage and energy used.

Electric Current

- In an electric circuit, charged electrons move through the wire and components.
- The rate of the flow of charge is called the electric current (I).
- Current is measured in amperes (A) using an ammeter.
- The battery in a circuit provides energy to the charged electrons passing through.
- The battery cell has a negative and positive terminal.
- The potential difference (p.d.) is the work done to move a unit of charge from one point to another in a circuit.
- Potential difference is measured in volts (V) using a voltmeter.
- In an electric circuit diagram the wires and components are drawn in a simple way to show the connections and components.
- A **series** circuit has a single loop:

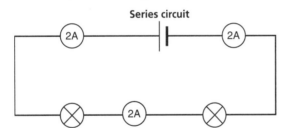

- In **parallel** circuits the flow of electric current is split between different branches. When the branches meet up, the currents add together again.

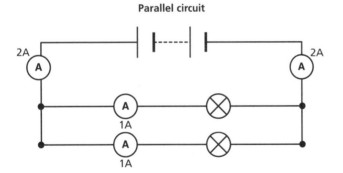

- Electricity in the home is connected as a parallel circuit, called a ring main.

> **Key Point**
>
> Potential difference is also known as voltage.

Ammeter

Voltmeter

Bulb

Cell

Components and Resistance

- All components in a circuit resist the flow of electric current; they have a **resistance**.
- An electrical **insulator** has very large resistance, allowing no current to flow through. A **conductor** has low resistance to the flow of current.
- Some components are engineered to be electrical insulators under certain conditions and conductors in others, e.g. light dependent resistors.
- Resistance is measured in ohms (Ω).
- The higher the resistance, the less the current flow through the component.
- A light bulb has a resistance, so electric charge can pass energy to the filament of the bulb, which produces light as it heats up.

> **Key Point**
>
> Resistance is the ratio of the p.d. to the current.

Electricity Calculations

- Resistance (R) can be calculated using the following formula:

$$R = \frac{V}{I}$$

- Where R = resistance (Ω), V = potential difference (V) and I = current (A).
- The current in a series circuit is the same all the way around the circuit.
- Electric power (P) is measured in watts (W) and is calculated using the formula:

$$P = V \times I$$

- Amount of energy transferred is calculated using the formula:

$$\text{energy transferred (kWh)} = \text{power (kW)} \times \text{time (h)}$$

> **Key Point**
>
> Kilowatts (kW, 1000 W) are used rather than watts because otherwise the numbers involved get large and difficult to handle very quickly.

Power Ratings

- All electrical equipment has a power rating (in W or kW) which enables you to work out how much electricity they use. The higher the power rating, the more electricity used.
- The time the equipment is switched on for is also important.
- A 2 kW heater switched on for 4 hours uses 2 kW × 4 h = 8 kWh
- Electricity bills charge for electricity based on how many kWh of electricity have been used.

Quick Test

1. What is the resistance if the p.d is 1.5 V and the current 3 A?
2. What is the unit for resistance?
3. Explain what an insulator is.
4. Draw a series and parallel circuit, each with two bulbs.

> **Key Words**
>
> series
> parallel
> resistance
> insulator
> conductor

Magnetism and Electricity

You must be able to:

- Describe magnetic attraction, repulsion and fields
- Explain the Earth's magnetic field and how it can be used for navigation
- Describe how electromagnets work.

Magnets

- Permanent magnets are made from magnetic metals and alloys.
- The three magnetic metals are iron, cobalt and nickel.
- Magnets have a North and South-seeking pole at each end.
- When two magnets are brought near one another, they will either attract or repel each other.

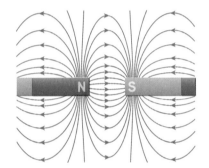 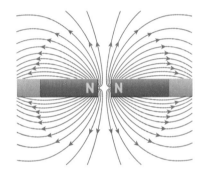

- The **magnetic field** of a magnet is invisible but can be shown using plotting compasses or iron filings sprinkled onto paper.
- The magnetic field can be represented by drawing **field lines**.

The Earth as a Magnet

- The Earth behaves like a bar magnet because it has a core made of iron and nickel.

Key Point

The closer the field lines, the more powerful the magnetic field.

Earth's magnetic field

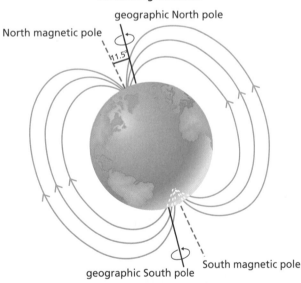

A compass needle points to the north because it is being attracted. Unlike poles attract so the North magnetic pole is actually a magnetic south pole.

- The poles of the Earth are the equivalent of the poles of a bar magnet.
- Navigation is possible using a handheld compass, which seeks magnetic North.

Electromagnetism

- When an electric current is passed through a wire, it causes a magnetic field to be formed.
- The strength of the magnetic field can be increased by:
  - coiling the wire
  - increasing the current flowing through the circuit
  - adding a core made from a magnetic metal, e.g. iron.
- As the magnetic field is temporary and created by an electric current, the magnet formed is called an **electromagnet**.
- Electromagnets are stronger than permanent magnets and can be turned off.

Magnetic fields through an electric coil

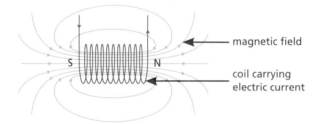

magnetic field

coil carrying electric current

Key Point
You could test the strength of an electromagnet by seeing how many paper clips it could pick up.

Uses of Electromagnets

- Electromagnets are used in a number of different ways, e.g. heavy lifting in car breakers yards.
- In a **DC motor**, the wire and core can move freely.
- Brushes enable the wires to make contact without tangling.
- The motor will continue to turn for as long as there is a current.

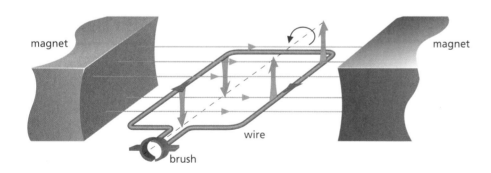

magnet

magnet

wire

brush

Quick Test

1. Name the ways an electromagnet's strength could be increased.
2. Suggest how the Earth's magnetic field is like a bar magnet.
3. What does it mean if the magnetic field lines are close together?
4. What are the three magnetic metals?

Key Words
magnetic field
field lines
electromagnet
DC motor

Waves and Energy Transfer

You must be able to:

- Explain how waves can be visualised using a ripple tank
- Explain how waves interact to produce interference
- Compare and contrast waves in sound and water with light waves.

Observing Waves

- It is possible to see the shapes of waves as they travel through water.
- The energy causes undulations which travel through the water.
- A ripple tank can be used to view the waves formed when a bar rapidly hits the water.
- When viewed from above the waves can be seen travelling with a **transverse** motion (at right angles to the direction of travel).
- The waves in water can also be reflected.

> **Key Point**
>
> Remember, the number of waves a second is the frequency. The more often waves appear, the higher the frequency.

Ripple tank with waves

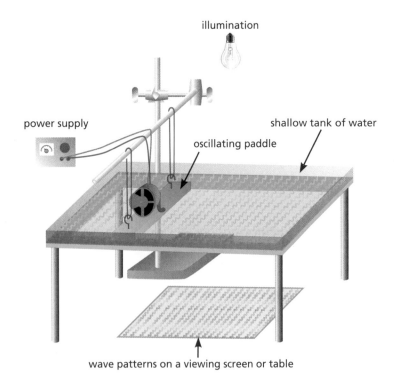

illumination

power supply

oscillating paddle

shallow tank of water

wave patterns on a viewing screen or table

- A wave viewed from the side has the following features:
  - The height of the wave is the **amplitude**
  - The top of the wave is the **peak**
  - The distance between a point on one wave and the same point on the next wave is the **wavelength**.

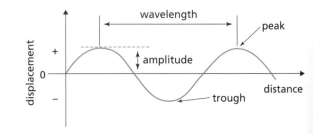

- Although the wave moves from left to right, a boat would move up and then down over the same spot.

Boat rotation on a wave

wave motion ⟶

Superposition

- Waves can interact with each other. This is called interference.
- If the peaks and troughs arrive at exactly the same time as the other wave then they will combine to produce waves that are the sum of each contributing wave.
- If the waves arrive out of phase, then they will cancel each other out.

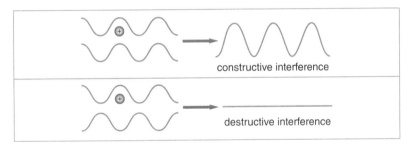

constructive interference

destructive interference

> **Key Point**
>
> All waves, including sound and light, can interact in this way.

Comparing Sound with Light

- Light travels in waves, but unlike sound or water waves, light does not need a medium (made of particles) to travel through.
- Light can travel in a vacuum (the absence of particles).
- The speed of light in a vacuum is 300,000 km/s.
- In other substances, such as air, water and plastic, the speed of light is slower.

> **Quick Test**
>
> 1. What makes light waves different to sound or water waves?
> 2. What device can be used to look at waves in water?
> 3. Approximately how many times faster does light travel than the speed of sound in air?
> 4. What happens when two waves arrive in phase?

> **Key Words**
>
> transverse
> amplitude
> peak
> interference

Waves and Energy Transfer

You must be able to:

- Explain how light is scattered and reflected off surfaces
- Understand the law of reflection
- Explain how white light can be split into the colours of the spectrum.

Light and Materials

- When light waves hit an object the light may be absorbed or reflected by the object.

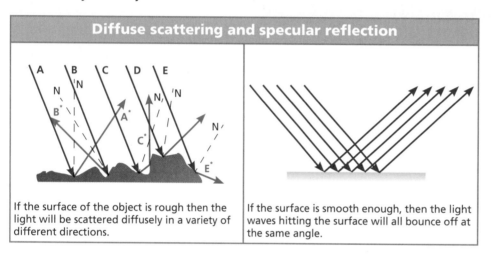

Diffuse scattering and specular reflection	
If the surface of the object is rough then the light will be scattered diffusely in a variety of different directions.	If the surface is smooth enough, then the light waves hitting the surface will all bounce off at the same angle.

Reflection

- Light waves can be drawn as rays.
- Light follows the law of **reflection**, which states that
  the angle of incidence = the angle of reflection
- An observer looking at the reflected rays will see an image.

Key Point

Incident and reflected rays are compared to what a normal ray of light would do. A normal is drawn at right angles to the surface.

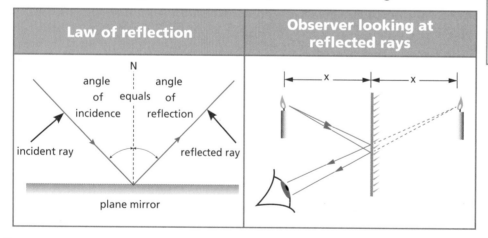

Law of reflection	Observer looking at reflected rays

The Eye

- The most simple camera is called a pinhole camera.
- It has no lens, just a small hole for light to pass through.

- The eye has a lens which focuses rays of light.
- The image forms on the retina at the back of the eye. The image information is sent to the brain and the image reversed, so that it is 'seen' the right way up.

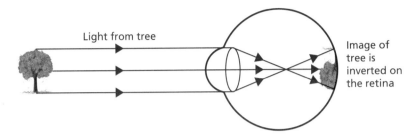

Light from tree

Image of tree is inverted on the retina

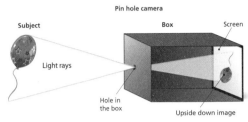

Pin hole camera

Subject

Box

Screen

Light rays

Hole in the box

Upside down image

- A **convex** lens brings the light to a focus on the retina by a process called **refraction**.
- Refraction is where light slows as it passes through an optically dense medium or speeds up going into an optically less dense medium.
- Refraction causes the ray to bend.

Seeing Colour

- Light waves carry energy.
- Specialised cells in the eye are hit by the light ray, causing a chemical reaction. An electrical signal is then sent to the brain.
- With a digital camera the light hits a sensor which then sends an electrical signal to the memory card.
- Visible light is made up of waves of different frequencies.
- A prism splits white light into the colours of the spectrum, based on their **frequency**.
- Objects absorb different frequencies of light.
- The light that is not absorbed is reflected, giving the object colour.

Focus of light on the retina

convex lens

focal point

Inverted image formed on retina

> ### Key Point
>
> Although we can see millions of different colours, our eyes only have receptors for red, green and blue.

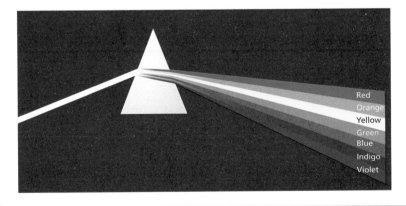

Red
Orange
Yellow
Green
Blue
Indigo
Violet

Key Words

reflection
convex
refraction
frequency

Motion On Earth and in Space

1 Which of the following examples of motion is the result of the forces acting on the vehicle being balanced?

a) accelerating to 80 km/h

b) decelerating to 20 km/h

c) braking to avoid an accident

d) travelling at 60 km/h in a straight line. [1]

2 The diagram below shows the planets in our solar system.

a) The planets all orbit the Sun.

What force keeps the planets in orbit? [1]

b) It takes light 4.1 h to travel from the Sun to Neptune.

Taking the speed of light to be 300,000 km/s, calculate the distance (in km) of Neptune from the Sun.

Show your working out. [3]

3 Explain what causes seasonal variation in the climate on the Earth. [3]

Energy Transfers and Sound

1 An elephant is being lifted using a lever.

a) Ollie is trying to work out the best way to lift the elephant with the lever. Which of the following suggestions would enable him to lift the elephant more easily?

   i) move the pivot further from the elephant   **ii)** stand closer to the pivot

   iii) move the pivot closer to the elephant   **iv)** stand further from the pivot   **[2]**

b) When the elephant is lifted it gains energy. Which of the following energy stores in the elephant will have been added to?

   i) sound   **ii)** light   **iii)** kinetic   **iv)** gravitational potential   **[1]**

2 An anti-loitering device was created to deter teenagers from hanging around shops. It emitted a loud and high pitched sound which only the teenagers could hear.

a) Suggest why teenagers could hear the anti-loitering device and why adults could not.   **[2]**

b) The device used a loudspeaker that could vibrate at very high frequencies. Explain how the sound travelled from the loudspeaker to a teenager's ear.   **[4]**

3 This diagram shows a pinhole camera.

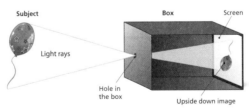

a) Describe how an image is formed on the screen.   **[1]**

b) Explain why the image is upside down.   **[1]**

c) The camera is now replaced with another camera, longer than the first one, but held in the same place. Describe what happens to:

   i) the size of the image   **[1]**

   ii) the brightness of the image.   **[1]**

Magnetism and Electricity

1. Draw a bar magnet marking the North and South poles and showing the magnetic field lines. [2]

2. Andy and Catherine are building a simple electromagnet.

   a) They want to increase the strength of the electromagnet. Which of the following would be ways to increase strength?

   i) more turns in the wire

   ii) use wire of a greater resistance

   iii) use thicker wire

   iv) add an iron core

   v) add a plastic core

   vi) increase the current [3]

   b) Suggest advantages of an electromagnet compared with a permanent magnet. [1]

3. The parallel circuit below shows four ammeters with two identical bulbs. If the first ammeter (A1) reads 4 A, what would each of the other ammeters read? [2]

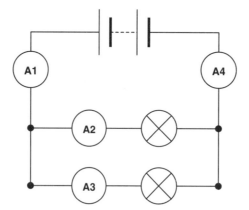

4. a) What current will flow through a 60 W bulb with a voltage of 230 V? [1]

   b) Calculate the energy (in kWh) transferred if the bulb was on for 200 hours.

   Show your working out. [3]

Waves and Energy Transfer

1 Naveen shines a ray of light at a mirror. She angles the ray of light so that it is at 30° to the normal.

Copy the diagram below and add the incident and reflected rays. [3]

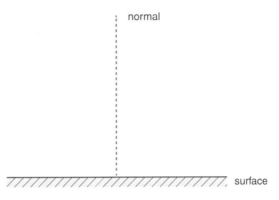

2 **a)** Copy the diagram below and add the labels **peak, trough** and **amplitude** to the waves shown. [3]

b) The wave is traveling from left to right. After one second, the second peak has reached the place the first peak was at. What will the **frequency** of the wave be?

Give the correct unit. [2]

3 White light is shone through a prism. What is the order of colours that appear on the screen? [4]

Magnetism and Electricity

1 Which of the following elements are magnetic?

a) sulfur   b) carbon   c) cobalt   d) vanadium   [1]

2 In which of the following circuits will current flow?   [1]

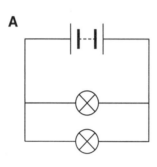

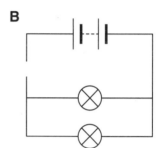

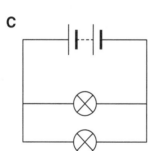

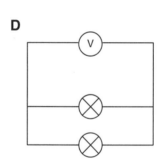

3 The diagram below shows a relay switch. It enables a small, safe circuit to be switched on and, in doing so, turn on a more dangerous circuit with much higher voltages.

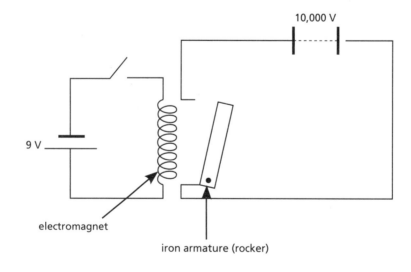

a) What will happen to the electromagnet when the switch is closed?   [1]

b) Explain what will happen to the armature once the switch is closed.   [2]

Waves and Energy Transfer

1 Explain why we can see our reflection in a mirror but not in a sheet of paper. [3]

2 This question is about sonar. Sonar waves are sound waves which can be sent out from a boat through the water and any reflections detected.

a) Explain how using sonar to locate fish is an example of echoes. [2]

b) Suggest how sonar could also be used to determine the depth of water beneath the boat. [2]

c) The boat emits a sonar wave and detects a reflection from the sea bed after 0.5s. The wave travels at 1500m/s. Calculate the depth of the sea at that point. [4]

3 Sara is looking at a tree.

a) Draw the light rays from the top and bottom of the tree and show their route through the eye. [3]

b) What is the name of the process by which light is bent by the lens? [1]

1. Quetzal is carrying out the distillation of alcohol. He sets up the apparatus as shown in the diagram below.

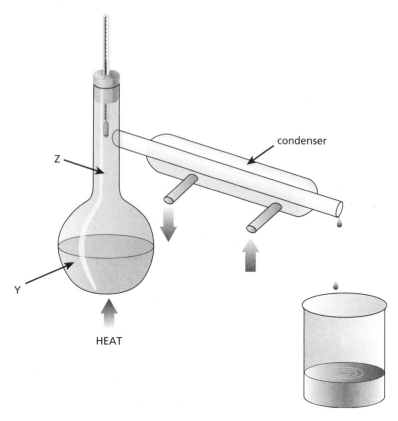

a) i) On the diagram above, write the letter B where the temperature will be the **highest** in the experiment.

1 mark

   ii) On the diagram above, label the distillate using the letter X.

1 mark

b) Suggest where cold water enters into the condenser and why it is important that the cold water flows in at that point.

....

....

....

....

....

2 marks

c) Ethanol is an alcohol.

It has the chemical formula C_2H_5OH

How many **atoms** are present in a molecule of ethanol?

Tick the correct box.

2 ☐   8 ☐

3 ☐   9 ☐

1 mark

d) There are a number of changes in state that take place during distillation.

In the boxes below draw the particle arrangement for the particles at point Y and at point Z.

Point Y   Point Z

1 mark

TOTAL

6

2 A group of performers are carrying out acrobatics for a TV talent show.

Michael and Kerrie lift a pole from each end and keep it level.

a) If Kerrie has to use 20 N of force to keep her end of the pole up, how much will Michael use to keep it level?

Tick the correct box.

80 N ☐

100 N ☐

20 N ☐

150 N ☐

☐
1 mark

b) Amy vaults onto the middle of the pole and does a handstand.

Amy weighs 600 N.

How much force will Michael have to use to keep balancing the pole with Kerrie?

Show your working out.

....

....

☐
2 marks

c) Nicola is wearing a pair of stilts.

Nicola has a weight of 650 N.

The end of each stilt has an area of 0.0009 m².

i) What **pressure** would the stilts exert on the ground? Give your answer in N/m².

Show your working out.

2 marks

ii) Describe what Nicola would need to do to exert **less** pressure on the ground.

1 mark

TOTAL

6

3 The lungs are where our body takes in oxygen from the air.

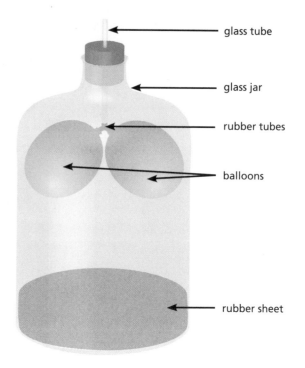

glass tube

glass jar

rubber tubes

balloons

rubber sheet

a) The diagram is a model that shows how the human lungs work.

Refer to the model to explain how it shows **breathing in** and **breathing out**.

Breathing in ....

....

....

....

3 marks

Breathing out ....

....

....

....

3 marks

b) The diagram shows the inside of three breathing airways.

One is **normal**, one is of a **heavy smoker**, and one is of someone with **asthma**.

Complete the label for each diagram.

Choose from these words: **normal**   **smoker**   **asthma**

i) ....   ii) ....   iii) ....

3 marks

c) Oxygen moves from the lungs to the blood stream by diffusion.

The diagram shows the relative concentrations of oxygen in three different cells.

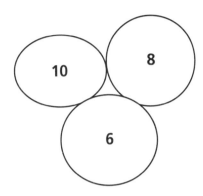

i) Draw three arrows (→) on the diagram to show the movement of glucose between the three different cells.

3 marks

ii) Explain why oxygen moves between cells in this way.

_____

_____

3 marks

d) Oxygen is used by the body during aerobic respiration.

There are two types of respiration in humans; aerobic and anaerobic.

i) Other than the use of oxygen, describe two differences between aerobic and anaerobic respiration in humans.

_____

_____

_____

2 marks

ii) Describe the difference between anaerobic respiration in humans and anaerobic respiration in yeast.

_____

_____

2 marks

TOTAL

19

4 Naveen is investigating how iron reacts with different chemicals.

In her first experiment she added an iron nail to three different boiling tubes to see what conditions are needed for rusting to take place.

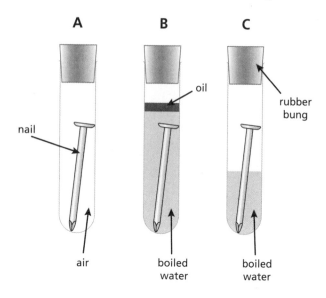

After a week she observed what had happened:

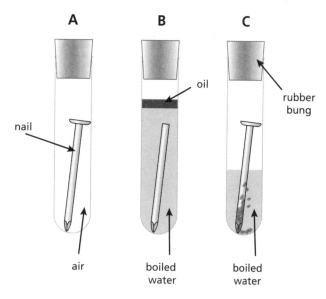

a) In one of the tubes a chemical reaction has taken place.

What evidence is there that a chemical change has taken place?

....

1 mark

b) Write the **word equation** for the rusting of iron.

.... + ... + ... ⟶ ....

2 marks

c) Write a conclusion for this experiment.

....

....

....

2 marks

d) Iron is more reactive than copper.

Naveen placed an iron nail into a test tube containing blue copper sulfate solution.

iron nail

blue copper
sulfate solution

i) Suggest what Naveen would **see** in this reaction.

....

....

....

2 marks

ii) Explain your answer.

....

....

....

2 marks

TOTAL

9

Mixed Test-Style Questions

5 This question is about plants and animals.

a) Look at the diagram of a food web.

foxes

blue tits

dormice

A farmer uses a pesticide to kill leaf-eating insects.

leaf eating insects

moths

Suggest what effect a reduction of leaf-eating insects would have on the numbers of **blue tits** and **moths**. Explain your answer.

oak tree

grass

Blue tits ....

....

2 marks

Moths ....

....

2 marks

b) Both plants and animals can respire, but only plants can photosynthesise.

Read the statements about photosynthesis and respiration. Some are true. Some are false.

Put a tick (✓) in the boxes next to the true statements.
Put a cross (✗) in the boxes next to the false statements.

	✓ or ✗
Respiration stores energy as light	
Photosynthesis uses energy from light	
Respiration breaks down large molecules to smaller molecules	
Photosynthesis creates organic molecules from inorganic molecules	
Respiration releases energy from chlorophyll	
Photosynthesis stores energy as light	

6 marks

c) Many species of plants and animals are in danger of extinction.

i) Red kites are a type of bird found in England. Some people say that red kites were once extinct in England.

Explain why this is the wrong use of the word extinction.

....

....

☐

1 mark

ii) Dinosaurs once existed. Now they are extinct.

Suggest what must happen to cause species such as dinosaurs to become extinct.

....

....

☐

2 marks

d) Animals and plants are interdependent.

Complete the diagram to show how animals and plants depend upon one another for **oxygen** and **carbon dioxide**. Write the name of each gas in the correct space.

plants

animals

☐

1 mark

e) Many scientists are responsible for our understanding of the interdependence of plants and animals.

Which of these statements best describe how scientists work?

Put ticks (✓) in the boxes next to the **three** best answers.

Most scientific discoveries are the result of a scientist building on the work of a previous scientist	
Once a scientist has an idea they never change their mind	
Scientists never share their ideas with others	
Scientists have their results checked by other scientists	
Scientists use data from experiments to check their ideas	
Scientists never consider risk when doing experiments	

☐

3 marks

TOTAL

☐

17

6 The diagram below shows Katie's toy train.

The pieces are made of wood, but have a magnet at each end. The magnet at one end of each carriage is separate to the magnet at the other end.

X   Y   Z

a) The magnet on carriage X repelled the magnet on carriage Y. The magnet on carriage Y repelled the magnet on carriage Z. The magnet on the right-hand end of carriage X is a North pole. The magnet on the left-hand end of carriage Z is a South pole.

    i) **On the diagram above**, label the poles on the magnets on each end of Carriage Y.

                                                            1 mark

    ii) Katie turned carriage Y around.

        The carriage X and carriage Z were **not** turned around.

        Describe what would happen now when Katie pushed the parts of the train together.

        Explain your answer.

                                                            2 marks

b) Katie's brother, Robert, has three metal objects. One is a bar magnet, one is a steel rod and one is an aluminium rod.

He takes one of the train carriages, brings each object close to the magnet on the carriage and observes what happens.

He records his observations in a table.

Complete Robert's table to show which object is being brought close to the magnet.

Test	Observation	Object
A	Attracts whichever side is pointing at the train carriage	
B	Attracts when in one position and repels in the opposite position	
C	Nothing happened	

2 marks

c) Which of the following are magnetic elements?

Tick the correct box(es).

Cobalt ☐

Steel ☐

Iron ☐

Vanadium ☐

1 mark

TOTAL ☐

6

Key

relative atomic mass
atomic symbol
name
atomic (proton) number

1	H hydrogen 1

Group 1	Group 2												Group 3	Group 4	Group 5	Group 6	Group 7	Group 0
																		4 **He** helium 2
7 **Li** lithium 3	9 **Be** beryllium 4												11 **B** boron 5	12 **C** carbon 6	14 **N** nitrogen 7	16 **O** oxygen 8	19 **F** fluorine 9	20 **Ne** neon 10
23 **Na** sodium 11	24 **Mg** magnesium 12												27 **Al** aluminium 13	28 **Si** silicon 14	31 **P** phosphorus 15	32 **S** sulfur 16	35.5 **Cl** chlorine 17	40 **Ar** argon 18
39 **K** potassium 19	40 **Ca** calcium 20	45 **Sc** scandium 21	48 **Ti** titanium 22	51 **V** vanadium 23	52 **Cr** chromium 24	55 **Mn** manganese 25	56 **Fe** iron 26	59 **Co** cobalt 27	59 **Ni** nickel 28	63.5 **Cu** copper 29	65 **Zn** zinc 30		70 **Ga** gallium 31	73 **Ge** germanium 32	75 **As** arsenic 33	79 **Se** selenium 34	80 **Br** bromine 35	84 **Kr** krypton 36
85 **Rb** rubidium 37	88 **Sr** strontium 38	89 **Y** yttrium 39	91 **Zr** zirconium 40	93 **Nb** niobium 41	96 **Mo** molybdenum 42	[98] **Tc** technetium 43	101 **Ru** ruthenium 44	103 **Rh** rhodium 45	106 **Pd** palladium 46	108 **Ag** silver 47	112 **Cd** cadmium 48		115 **In** indium 49	119 **Sn** tin 50	122 **Sb** antimony 51	128 **Te** tellurium 52	127 **I** iodine 53	131 **Xe** xenon 54
133 **Cs** caesium 55	137 **Ba** barium 56	139 **La*** lanthanum 57	178 **Hf** hafnium 72	181 **Ta** tantalum 73	184 **W** tungsten 74	186 **Re** rhenium 75	190 **Os** osmium 76	192 **Ir** iridium 77	195 **Pt** platinum 78	197 **Au** gold 79	201 **Hg** mercury 80		204 **Tl** thallium 81	207 **Pb** lead 82	209 **Bi** bismuth 83	[209] **Po** polonium 84	[210] **At** astatine 85	[222] **Rn** radon 86
[223] **Fr** francium 87	[226] **Ra** radium 88	[227] **Ac*** actinium 89	[261] **Rf** rutherfordium 104	[262] **Db** dubnium 105	[266] **Sg** seaborgium 106	[264] **Bh** bohrium 107	[277] **Hs** hassium 108	[268] **Mt** meitnerium 109	[271] **Ds** darmstadtium 110	[272] **Rg** roentgenium 111								

Elements with atomic numbers 112–116 have been reported but not fully authenticated

*The Lanthanides (atomic numbers 58–71) and the Actinides (atomic numbers 90–103) have been omitted.

Cu and **Cl** have not been rounded to the nearest whole number.

Answers

Pages 4–5 **Review Questions**

Page 4

1. a) nutrition; growth; reproduction [3]

MRS GREN - This mnemonic can help you remember the characteristics of living things (Movement, Respiration, Sensitivity, Growth, Reproduction, Excretion, Nutrition).

b)

Characteristic	Definition: what the word means
Respiration	Converting energy from carbohydrates and fats into energy.
Excretion	Getting rid of waste products. **[1]**
Sensitivity	Being able to sense and respond to changes in the environment. **[1]**
Nutrition	Getting nutrients by eating other organisms or from their environment. **[1]**

Page 5

2. a) i; b) ii; c) i or iii; d) ii; e) ii or iii [5]

One material may have several different properties.

3. a) iii [1];
   b) i; iii [1 mark for both correct]
   c) the Earth orbits the Sun – 365 days; the Earth rotates once – 1 day; the Moon orbits the Earth – 28 days [3]

Draw straight lines using a ruler.

Pages 6–11 **Revise Questions**

Page 7 Quick Test

4. cell wall, (large, permanent) vacuole, or chloroplast
5. mitochondria
6. diffusion
7. cell, tissue, organ, system, organism

Page 9 Quick Test

1. sperm and egg
2. day 14
3. Insect pollinated — flower is brightly coloured, produces less pollen, produces nectar to attract insects. Accept reverse argument for wind pollinated flower.
4. animal, wind or self-dispersal

Page 11 Quick Test

1. The ribs move up and outward while diaphragm moves downwards.
2. volume decreases, pressure increases
3. diffusion
4. Smoking stops the mechanism for getting rid of mucus. Mucus builds up and causes coughing. Lung infections are more likely and there is a long term risk of cancer.

Page 13 Quick Test

1. **Five from:** Protein, fat, carbohydrate, vitamins, minerals, fibre, water.
2. Obesity, starvation, kwashiorkor, mineral deficiency, e.g. anaemia, vitamin deficiency, e.g. scurvy.
3. Mouth, oesophagus, stomach, small intestine, large intestine, rectum, anus.
4. Plants make their own food (photosynthesis), animals consume food.

Pages 14–15 **Practice Questions**

Page 14

1.

Membrane	Controls what enters and leaves a cell
Cytoplasm	Where chemical reactions take place
Nucleus	Stores information and controls the cell
Mitochondria	Releases energy from glucose
Cell wall	Supports the cell
Vacuole	Keeps the cell firm
Chloroplast	Uses light energy to produce food

[1 mark each up to a maximum of 7 marks]

Draw straight lines using a ruler.

2.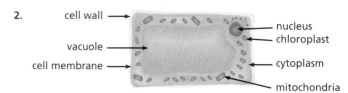

[1 mark for each label, up to a maximum of 6 marks]

3. iii [1]

Page 15

1. a) A diet that contains the right balance [1] and right amounts [1] of all the nutrients needed.
   b) Carbohydrates – to provide energy [1]
      Proteins – for growth and repair [1]
      Fibre – helps undigested food/waste products pass through the body [1]

2. a)

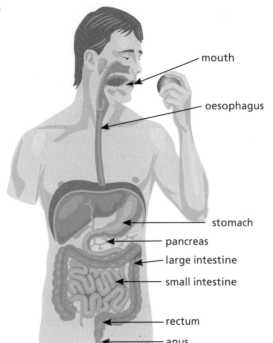

- mouth
- oesophagus
- stomach
- pancreas
- large intestine
- small intestine
- rectum
- anus **[8]**

b) a) mouth – food broken into pieces
   b) oesophagus – food passes from mouth to stomach
   c) stomach – food mixed with acid and enzymes to help break it down
   d) pancreas – produces enzymes to break down food
   e) large intestine – water absorbed into blood stream
   f) small intestine – broken down food absorbed into blood stream
   g) rectum – waste material stored
   h) anus – waste material eliminated **[8]**
3. feeding in animals involves eating food **[1]** that is broken down **[1]**; feeding in plants involves making food **[1]** from simple substances (water and carbon dioxide), by photosynthesis **[2]**

> State clearly whether you are writing about plants or animals.

Pages 16–23 **Revise Questions**

Page 17 Quick Test
1. aerobic
2. aerobic
3. anaerobic
4. anaerobic
5. Because he cannot take in enough oxygen for aerobic respiration and so anaerobic respiration takes place, producing lactic acid which causes the muscles to feel tired and painful.

Page 19 Quick Test
1. support, protection, enables us to move, makes red blood cells
2. ligaments
3. tendons
4. Working against each other or in opposite directions.

Page 21 Quick Test

1. water + carbon dioxide $\xrightarrow[\text{chlorophyll}]{\text{light}}$ glucose + oxygen

2. Respiration uses glucose and oxygen to release energy, and water and carbon dioxide are produced. In photosynthesis, carbon dioxide and water are converted into glucose and oxygen using light energy.
3. Because the leaves are the site of photosynthesis and this involves the movement of carbon dioxide into the leaf/oxygen out of the leaf through the stomata.
4. Plants make food and oxygen for animals. Animals make carbon dioxide for plants.
5. How different organisms rely upon each other for their survival.

Page 23 Quick Test
1. How energy moves through the food web.
2. Build-up of poisons towards the top of a food chain.
3. Variation is the differences between organisms. It increases the chances of survival when the environment changes.

Pages 24–25 **Review Questions**

Page 24
1. unicellular organisms need different structures to enable them to survive **[1]**; in different environments **[1]**

> Look for key words. The key word in this question is 'explain'.

2. bone cell – skeletal system **[1]**
   red blood cell – transport system **[1]**
   nerve cell – nervous system **[1]**
   sperm cell – reproductive system **[1]**

> Draw straight lines using a ruler.

3.

Part	Male	Female
Testis	✓	✗
Egg cell	✗	✓
Vagina	✗	✓
Sperm	✓	✗
Penis	✓	✗

[5]

4. insect pollinated flowers are brightly coloured **[1]**, have a scent **[1]** and produce nectar to attract insects **[1]**, *or* wind pollinated flowers are not designed to attract insects so are not brightly coloured **[1]**, do not have a scent **[1]** or produce nectar but do produce lots of pollen **[1]**.

[total of 3 marks]

> Show clearly if your answer relates to insect or wind pollinated flowers.

5. They pollinate our crops **[1]** and without pollination there would be no crops to harvest **[1]**
6. Place pollen grains on a microscope slide. **[1]**
   Cover pollen grains with a cover slip. **[1]**
   Place slide on the microscope stage. **[1]**
   Switch on lamp or adjust mirror. **[1]**
   Select lens for suitable magnification. **[1]**
   Focus the image with the focusing knob. **[1]**

Page 25
1. When we breathe in, the ribs move up and out and the diaphragm moves down **[1]**. Pressure drops **[1]** and volume increases **[1]**. When we breathe out, the ribs move in and down and the diaphragm moves up **[1]**. Pressure increases **[1]** and volume decreases **[1]**.
2. **Any two from**: It increases the strength of the diaphragm and intercostal muscles; It increases the vital lung capacity/the volume of air that can be forcibly exhaled after inhaling fully; It improves gas exchange **[2]**
3. carbohydrates – provide energy **[1]**
   fat – stores energy **[1]**
   proteins – used to grow new cells **[1]**
   vitamins – needed for chemical reactions to take place **[1]**
   minerals – needed for strong bones and blood **[1]**
   fibre – speeds movement up through the gut **[1]**
   water – dissolves chemicals for chemical reactions to take place **[1]**
4. Bacteria help break down food **[1]**, so it can be absorbed into the blood **[1]**.
5. They help break down the food we eat **[1]**, by speeding up chemical reactions **[1]**.

> Learn definitions. They are an easy way to score marks.

6. The small intestine is where water is absorbed. **[1]**

Page 26

1. uses oxygen – aerobic [1]
   produces lactic acid – anaerobic [1]
   produces alcohol – anaerobic [1]
   releases the most energy – aerobic [1]
   fermentation – anaerobic [1]

   > Only write one tick in each row or you will not be given the marks.

2. In yeast it produces carbon dioxide [1], and alcohol [1]. In humans it produces lactic acid [1].
3. The skeleton provides support for the body [1], enables movement to occur because of the joints [1], protects the brain heart and lungs [1], and makes red blood cells in the long bones [1].
4. 1 – skull; 2 – ribs; 3 – spine; 4 – elbow joint; 5 – hip joint; 6 – femur; 7 – kneecap; 8 – radius and ulna; 9 – pelvis; 10 – humerus [10]
5. Muscles can only contract [1], so two muscles are needed to move the joint in opposite directions [1]. This is called an antagonistic pair [1].

   > Remember, muscles can only contract, they cannot expand.

Page 27

1. produces oxygen – photosynthesis [1]
   produces carbon dioxide – respiration [1]
   uses energy from sunlight – photosynthesis [1]
   releases energy – respiration [1]
   requires chlorophyll – photosynthesis [1]

2.

Structural adaptation	Explanation
Leaves have a large surface area.	To catch as much light as possible.
Leaf has tiny holes called stomata.	To absorb carbon dioxide and release oxygen. [1]
Leaves have xylem tubes.	To carry water to the leaf. [1]
Leaf cells near top of leaf contain lots of chloroplasts.	The chloroplasts contain chlorophyll, a chemical needed for photosynthesis. [1]

[5]

3. a) Oak tree → leaf eating insects → blue tit → fox [1]
   Oak tree → moth → dormouse → fox [1]
   b) The oak tree [1]

   > Always start with the plants and draw the arrows pointing away from them.

4. pollinate flowers [1]; may be a pest that destroy crops [1].
5. Bioaccumulation [1] occurs because large animals eat lots of small ones so the poison builds up [1].

Page 29 Quick Test

1. 23 from mum and 23 from dad
2. chromosome
3. A single section of DNA that codes for protein
4. James Watson, Francis Crick, Maurice Wilkins, Rosalind Franklin

Page 31 Quick Test

1. Continuous, e.g. height, where a range exists between very short and very tall with most being somewhere near the middle of the range. Discontinuous, e.g. blood groups, where distinct groups exist.
2. sexual reproduction
3. The range of different organisms in an ecosystem.
4. A place where genetic material is collected and stored.

Page 33 Quick Test

1. A substance that affects the human body.
2. An unwanted effect on the human body.
3. Any two from caffeine, nicotine and alcohol.
4. The need to keep taking the drug and when they stop they get withdrawal symptoms.

Page 35 Quick Test

1. any three from: eyes produce tears that contain a chemical to kill microbes; ears produce wax to trap microbes; nose and throat produce mucus to trap microbes; skin acts as a barrier to microbes; stomach produces acid to kill microbes in food; urine flushes out microbes that enter the genitals; vagina produces acid to kill microbes
2. bacteria; viruses; fungi
3. when dead microbes are injected into the body causing the blood to make memory cells
4. antibiotics do not work against viruses

Page 36

1. Aerobic respiration uses oxygen [1]. Anaerobic respiration produces lactic acid [1], produces alcohol [1], releases the least energy [1].
2. a) glucose; carbon dioxide [2]
   b) carbon dioxide; alcohol (can be either way round) [2]
   c) lactic acid [1]
3. Respiration releases energy [1] for all the chemical processes living things need [1].

Page 37

4. supports the body; helps with movement; protects some organs; makes red blood cells. [4]

   > This is a two mark question so you need to write two parts to your answer.

5. 1 = cartilage; 2 = lubricating fluid;
   3 = ligament; 4 = tibia; 5 = femur [5]
6. a) contracts; bends arm [2]
   b) contracts; straightens arm [2]
   c) **One from:** antagonistic; muscles [1]
1. a) carbon dioxide; oxygen [2]
   b) light; chlorophyll [2]
2. animals make carbon dioxide; plants use carbon dioxide; animals use oxygen; plants make oxygen [4]

Page 38

1. 46; 46; 23; 23; 46 [5]
2. A – cell, B – nucleus, C – chromosome, D – gene [4]

Page 39

3. The graph shows a gradual range of values which is characteristic of continuous variation [1] A graph showing discontinuous variation would show discrete groups of variation [1].

1.

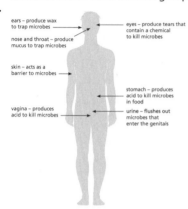

ears – produce wax to trap microbes
eyes – produce tears that contain a chemical to kill microbes
nose and throat – produce mucus to trap microbes
skin – acts as a barrier to microbes
stomach – produces acid to kill microbes in food
vagina – produces acid to kill microbes
urine – flushes out microbes that enter the genitals

[7]

2. A medical drug is used to treat disease [1]. A recreational drug is used for pleasure [1].
3. A depressant relaxes the body [1], e.g. cannabis or heroin [1]. A stimulant makes you alert [1], e.g. cocaine or amphetamine [1]. A hallucinogen alters the perception of reality [1], e.g. LSD or magic mushrooms [1].
4. Drinking can cause liver failure [1] and increased risk of heart failure [1]. Smoking increases the risk of heart disease [1] lung cancer [1] and lung infections [1].

Page 41 Quick Test

1. Filter the mixture using filter paper. The residue in the filter paper is the solid.
2. small
3. The evaporated and re-condensed liquid following boiling a mixture.
4. Draw a line approximately at the bottom (e.g. 0.5 cm from the bottom) of the chromatography paper in pencil. Place a sample of each liquid mixture on the line. Place the paper into a beaker containing the solvent (e.g. water). As the solvent moves up the paper the more soluble pigments move further. Once the first pigment has reached the top, remove the paper, mark where the solvent has reached using a pencil and then dry the paper.

Page 43 Quick Test

1.

Solid | Liquid | Gas

2. Calcium carbonate decomposes into calcium oxide and carbon dioxide when it is heated.
3. Oxidation is when substances gain oxygen in a reaction.
4. Water and oxygen. The process is faster if salt is also present.

Page 45 Quick Test

1. CO_2 and H_2O
2. The number of protons in the nucleus of an atom of the element.
3. two
4. three

Page 47 Quick Test

1. calcium + oxygen → calcium oxide
2. zinc sulfate
3. Three from: sonorous, shiny, malleable, conducts heat, conducts electricity, ductile, high melting point.
4. Three from: insulator (heat and electric), low melting points, gases at room temperature, low density.

Page 48

1. a) discontinuous [1]
   b) blood groups, tongue rolling [2]

   Your teacher may give you other examples.

2. when the environment is changing very quickly [1]
3. a) Watson and Crick produced a theory to explain the structure of DNA [1]
   b) Rosalind Franklin produced evidence to support their theory [1]

Page 49

1. pathogen – disease-causing microbe [1]
   toxin – poison produced by a disease-causing microbe [1]
   antibody – chemical produced by white blood cells to kill pathogens [1]
2. bacteria – tuberculosis [1]
   virus – polio [1]
   fungus – athlete's foot [1]
   [1 mark for each named microbe with a disease caused by it]

   Your teacher may give you other examples.

3. less addictive [1]
   less serious side effects [1]
4. Addiction is the need to take more and more of a drug [1]. Withdrawal is the physical effects on the body of not taking the drug [1].

   Learn definitions. They are an easy way to score marks.

5. Drinking alcohol – liver failure [1]
   Smoking – lung cancer [1]
   Taking LSD – trying to fly off a tall building [1]
   Using heroin – reduces breathing [1]
6. some white cells can engulf microbes [1]; some white cells produce chemicals to destroy microbes [1]; some white cells called memory cells protect us from future infections [1]

Page 50

1. a) distillation [1]
   b) chromatography [1]
   c) filtration [1]
2. a) A and C [2]

   Remember that elements that are gases at room temperature and pressure often travel round as pairs.

   b) C [1]
3. Pour the sea water mixture into a filter funnel [1]. Allow the solution to pass through [1]. Discard the residue and collect the filtrate [1]. Place the filtrate into an evaporating basin and heat [1]. Once the water has evaporated salt crystals will be left behind [1].

   The key here is to say what will happen in steps. A little like writing a recipe.

4. Vinnie could measure the distance each pigment moved on the chromatogram from the starting line using a ruler [1].

5. The filter removes solid particles/waste/food that are building up in the water [1]. Filtration works by passing a liquid containing solid particles through a barrier that only allows small molecules through [1], leaving the larger particles behind [1].

Page 51
1. a) iii [1]
   b) ii [1]
2. ii; iii [2]
3. a) B [1]
   b) C [1]
   c) A [1]
4. An element is made up of only one kind of atom [1]. A compound is made up of more than one different type of atom [1] chemically joined together [1].

Pages 52–59 **Revise Questions**

Page 53 Quick Test
1. In solids, the particles are closer together than in gases, so the density is greater.
2. Absolute zero (0 K) or –273 °C.
3. Sublimation is where a substance changes from a solid to a gas with no liquid phase.
4. Solids have all their particles touching and vibrating in position and have a defined shape. Liquids have at least 50% particles touching and move relative to each other and they take the shape of the container that they are in. Gases have no particles touching and will occupy all of the available space in a container.

Page 55 Quick Test
1. If the air in a sealed balloon is heated the gas particles will gain kinetic energy and move faster. This increases the air pressure, so the balloon will get bigger until the air pressure inside equals the air pressure outside. If the pressure becomes too great, the balloon may burst.
2. If the air in a sealed balloon is cooled the gas particles will lose kinetic energy and move more slowly. This decreases air pressure, so the balloon will get smaller until the pressure inside equals the air pressure outside.
3. Unlike other solids, the particles in ice are further apart than in a liquid. This means that there are fewer ice molecules in a given volume compared to the same volume of liquid water. Ice is less dense, so it will float.
4. Diffusion occurs where particles at a higher concentration move randomly to areas where they are present in a lower concentration. The bigger the difference, the faster the rate of diffusion.

Page 57 Quick Test
1. A catalyst increases the rate of reaction but is not used up in the reaction.
2. Activation energy is the minimum amount of energy needed to ensure that a reaction takes place.
3. A word equation just tells you what the reactants and products are, whereas a chemical equation tells you what atoms are present and the ratio of reactants and products.
4. (s), (l), (g), (aq)

Page 59 Quick Test
1. The substance is a weak acid.
2. Neutralisation is where an acid reacts completely with a base.
3. metal + acid ➔ salt + hydrogen
4. UI paper relies on using your eyes and judging the colour to find the pH, a pH probe and a data logger measures the actual pH. So pH probe and a data logger is both more accurate and more reliable. It is also more precise and it can be used over long periods of time.

Pages 60–61 **Review Questions**

Page 60
1. a) sample A = 3; sample B = 2 [2]
   b) sample A [1]
   c) The line is drawn in pencil because otherwise the line would move up the paper [1] along with the pigments being tested [1].
2. a) thermal decomposition [1]
   b) calcium carbonate ➔ calcium oxide + carbon dioxide [2]
3. $C(s) + O_2(g) \rightarrow CO_2(g)$ [3]

[3 marks. 1 mark for correct reactants on left-hand side, 1 mark for correct product on right-hand side, and 1 mark for correct state symbols.]

Page 61
1. a) The circles (atoms) have been rearranged [1] and joined together in a new way [1].
   b) Substance Y is hydrogen [1]. Substance Z is water/dihydrogen oxide/H_2O [1].
   c) Mass has been conserved in the reaction because the number of atoms of each element is the same on both sides of the equation. [1]
2. A = sulfuric acid; B = magnesium nitrate; C = hydrochloric acid; D = hydrogen; E = lead [4]
3. a) P;
   b) C;
   c) P;
   d) P;
   e) C [1]
4. c) $2Mg(s) + O_2(g) \rightarrow 2MgO(s)$ [1]

Although d) is balanced, it is incorrect as oxygen exists as O_2 molecules.

Pages 62–63 **Practice Questions**

Page 62
1. Solid should have at least 3 rows of atoms, all touching, with no gaps [1]. Liquid should have approximately 11 atoms, at least 50% of them touching, with gaps present [1]. Gas should have no atoms touching and show a random arrangement [1].
2. a) Sally would see blue air freshener appearing. [1]
   b) The air freshener is turning back into a solid from a gas as it cools. [1]
3. Brownian motion is the process by which large particles are hit by the random collision of air or water particles [1] causing the large particles to move in random directions [1].
4. When liquid, water molecules can fit closer together [1] than when they are arranged as a solid lattice structure [1]. This means that the ice is less dense than the liquid form of water [1].
5. b) The CO_2 gas particles are slowed down rapidly causing a solid to form.

Page 63
1. a)
[2]
   b) copper oxide [1]
2. c [1]
3. metal + acid ➔ **salt** + **hydrogen** [1]
4. a) red [1]
   b) orange / yellow [1]
   c) green [1]
   d) blue / purple [1]
5. a) magnesium chloride + hydrogen [1]
   b) copper nitrate + water [1]
   c) vanadium sulfate + water + carbon dioxide [1]

Pages 64–71 Revise Questions

Page 65 Quick Test
1. no
2. magnesium + iron sulfate → magnesium sulfate + iron
3. anything above sodium in the reactivity series, e.g. potassium.
4. The reactivity series is a list of the comparative reactivity of metals and carbon. It can be used to decide whether a displacement reaction will take place or not.

Page 67 Quick Test
1. They are unreactive, so when found would have been in their elemental form.
2. cool the clay very quickly
3. A combination of materials that, when bonded together, have new, enhanced properties.
4. chemically join monomer units together

Page 69 Quick Test
1. igneous, metamorphic and sedimentary
2. If the chemicals in the Earth's outer layer were soluble, the rain would have washed them away millions of years ago.
3. Check your answer against the picture on page 69.

Page 71 Quick Test
1. human activity, such as burning fossil fuels.
2. Rare or valuable parts of a product are extracted from waste and then used in a new product or process.
3.

4. Scientists believe that humans are causing climate change through activities such as burning fossil fuels, cutting down rainforests, livestock farming, and paddy fields. The carbon dioxide released is causing the climate to heat up (the greenhouse effect).

Pages 72–73 Review Questions

Page 72
1. a) A [1]
   b) B [1]
   c) C [1]
2. a) The water molecules are closer together when a liquid compared to when they are a solid [1]. This means that liquid water is denser than ice [1]. The liquid water at the bottom of the pond will not freeze [1].
   b) If Hilary switches the pump back on, the water will circulate [1] and so the water at the bottom will get colder [1] and possibly freeze [1].

Page 73
1. a) sodium + chlorine → sodium chloride
   [1 mark for reactants, 1 mark for products]
   b) $2Na(s) + Cl_2(g) \rightarrow 2NaCl(s)$ [2]
2. a) hydrogen peroxide → water + oxygen [2]
   b) a catalyst [1]

3. a)

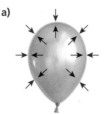
[2]

b) Particles should be well spaced out and randomly arranged. [1]

Pages 74–75 Practice Questions

Page 74
1. a) iron and tin [2]
   b) carbon + iron oxide → **carbon dioxide + iron** [2]
2. monomers [1]
3. Basalt is an igneous rock that cools very quickly [1].
4. A composite is a substance that consists of different materials bonded together [1]. Has different properties to the individual substances alone. [1]

Page 75
1.

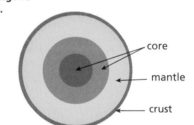
[3]

2.

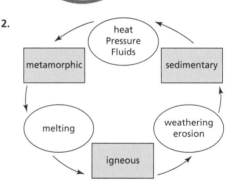
[5]

3. a) unusual changes in normal weather patterns [1]
   b) ii and iv [2]
4. Recycling resources is important as the resources are in limited supply [1]. Recycling means that the resources are available for use in other products [1]. **Also allow:** Less energy required to recycle old materials than to obtain new ones from their raw materials / Less carbon dioxide is produced by recycling.

Pages 76–83 Revise Questions

Page 77 Quick Test
1. unbalanced
2. the Newton (N)
3. Force arrows indicate the direction and strength of a force.
4. force × distance in a given direction; or moment is the turning effect of the force

Page 79 Quick Test
1. m/s or km/h [or any other suitable answer]
2. Work is when an object changes speed or shape.
3. 1.2 km/min or 0.02 km/s or 72 km/h or 20 m/s

Page 81 Quick Test

1. Mass is the amount of particles that make up an object (measured in kg). Weight is a force on the object due to gravity (measured in N).
2. The Earth has gravitational field strength due to its mass. The further away from the Earth's centre of mass, the less it will affect an object.
3. An electric field.
4.

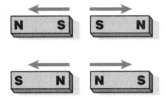

repelling   attracting

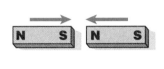

Page 83 Quick Test

1. Atmospheric pressure is due to the weight of air molecules pushing down on the surface. The higher you go, the fewer air molecules there are, therefore less to push down and so less pressure.
2. N/m^2
3. 1000 N
4. The upthrust from the water.

Pages 84–85 **Review Questions**

Page 84

1. gold or iron [1]
2. a) ii; iv [2]
   b) ii) iron + copper sulfate → iron sulfate + copper [1]
   iv) copper + silver nitrate → copper nitrate + silver [1]
3. Iron ore is heated at a high temperature with carbon [1]. The carbon displaces the iron, which pours out of the bottom of the furnace [1]. The carbon reacts with oxygen to form waste carbon dioxide [1].

Page 85

1.

Rock Type	Description
Igneous	is very hard and made of lots of small crystals
Sedimentary	may contain fossils
Metamorphic	often have a striped appearance and can be used to make statues

[3]

2.

Element	Abundance (%)
Oxygen	46
Silicon	28
Aluminium	8.0
Iron	5.0

[4]

3. a) Arwen and Ruth [1]
   b) Recycling does use energy [1]. However, the materials are rare and there would be a greater cost in trying to find more and extracting them [1]. It is really a question of using less energy [1]. [2 of the 3 possible marks]

Pages 86–87 **Practice Questions**

Page 86

1. a) Newtons
   b) 4 N
   c) Gravity/the weight of the apple [1]
2. a) 200 g [1]
   b) 150 g [1]
3. 60 s + 48 s = 108 s,
   800 m/108 s = 7.4 m/s [2]
4. 15 N x 0.1 m = 1.5 Nm [3]

Page 87

1. a) S N [2]
   b) Turn one of the magnets the other way round [1] so the poles are opposite to one another [1].
2. 10 kg – 100 N, 15.5 kg – 155 N, 2000 g – 20 N [3]

   Watch out for the units. The last mass is in g not kg.

3. a)

[1]

   b) The pressure on the water is greatest for the lowest hole. The pressure decreases towards the top of the bottle [1], so B will have less pressure than C, but more than A. A has the least pressure [1].
4. Amy 600 ÷ 0.025 m^2 = 24000 N/m^2 (or 2.4 N/cm^2)
   George 600 ÷ 0.035 m^2 = 17143 N/m^2 (or 1.7 N/cm^2) [2]

Pages 88–95 **Revise Questions**

Page 89 Quick Test

1. 1 N upwards
2. A distance–time graph can tell us how far an object travelled over a set time. This also enables speed to be calculated.
3. 130 mph
4. Whether the object is accelerating or decelerating slowly or rapidly.

Page 91 Quick Test

1. 12 h
2. because the Earth is tilted on its axis so the hemispheres receive different amounts of radiation from the Sun.
3. 2000 N
4. Mercury must be smaller in mass and so have lower gravitational field strength.

Page 93 Quick Test

1. fuel + oxygen → carbon dioxide + water + energy
2. Combustion is a much faster reaction, with the energy released producing heat and/or light. Respiration is very slow in comparison.
3. Radiation is the transfer of heat via waves (infrared).
4. 1000 J

Page 95 Quick Test

1. 50 Hz is 50 waves per second
2. An echo is a reflected sound wave.
3. To reduce echoes, place angled materials (like egg boxes) on the walls so that the waves from a singer or instrument are not reflected back.
4. A loudspeaker converts music to an electrical signal which causes electromagnets to move a fabric skin. This movement makes air particles move as sound waves which then reach the ear.

Pages 96–97 **Review Questions**

Page 96

1. a) no [1]
   b) yes [1]
   c) no [1]
2. 600 g – stretch was 12 cm; [1]
   700 g – stretch was 14 cm [1]
3. Speed = distance ÷ time
   27600 km/h = 42927 km ÷ time [1]
   time = 42927 km ÷ 27600 km/h [1]
   = 1.56 h (or 93.3 min) [1]

Page 97

1. c and d [2]
2. a) The rod has been charged with electrons / had electrons removed by rubbing **[1]** and the water, which must have the opposite charge, is attracted to it **[1]**.
   b) it will be repelled (as it has the same charge). [1]
3. a) **Any two from:** tyres; brake blocks; handlebar grips; pedal surfaces [2]
   b) **Any two from:** axles; chain links; steering [2]

Pages 98–99 **Practice Questions**

Page 98

1. A – Lisa jogged for 4 min; B – Lisa stopped for 2 min; C – Lisa walked for 8 min; D – Lisa stopped for 3 min [4]
2. 80 km/h [1]
3. a) an astronomical unit of distance, the distance that light travels in an Earth year [1]
   b) 1 light year = 300,000 km/s x 60 x 60 x 24 x 365
      = 9,460,800,000,000 km [1]
      4.2 light years = 9,460,800,000,000 km x 4.2 [1]
      = 39,735,360,000,000 km [1]
   Accept answer rounded to 40,000,000,000,000 km
4.

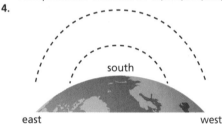

east   south   west [2]

Page 99

1. a) Just when cord goes taut. [1]
   b) As he steps off the bridge, but before he starts falling. [1]
2. a) A = oxygen; B = carbon dioxide; C = water [2 marks for all 3 correct]
   b) respiration [1]
3. c) radiation [1]
4. Jamie and Linda can hear each other because the sound waves are reflected **[1]** by the walls and also spread in all directions from their voices **[1]**.

Pages 100–107 **Revise Questions**

Page 101 Quick Test

1. R = V/I so R = 1.5 V/3 A = 0.5 Ω
2. the ohm (Ω)
3. An insulator is a substance that offers complete resistance to electric current. It does not conduct electricity.
4.
Series circuit

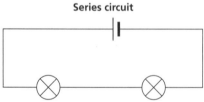

Parallel circuit

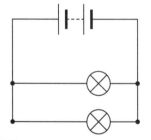

Page 103 Quick Test

1. increase number of coils, increase current, add an iron core
2. The Earth is like a bar magnet because it creates a North and a South pole and the field lines around the Earth resemble those of a bar magnet.
3. The closer the magnetic field lines, the stronger the magnetic field.
4. cobalt, iron and nickel

Page 105 Quick Test

1. light waves do not need a medium to travel through, so they can travel much faster
2. a ripple tank
3. light = 300,000,000 m/s and sound = 300 m/s so light travels approximately 1,000,000 times faster
4. Waves in phase add together, so increase in size (amplitude).

Page 107 Quick Test

1. violet
2. a prism
3. Light enters into the pinhole and forms an upside down image on the back of the camera. This is where the film would be placed if the image needed to be recorded.
4. where the light hits an uneven surface and reflects in different directions

Pages 108–109 **Review Questions**

Page 108

1. d) travelling at 60 km/h in a straight line [1]
2. a) gravity / gravitational pull from the Sun [1]
   b) 300,000 km/s x 60 x 60 = 1,080,000,000 km/h
      4.1 h x 1,080,000,000 km/h = 4,428,000,000 km [3]
3. The Earth has seasons because the planet is tilted **[1]**. This means that, as the Earth goes around the Sun the summer is when a hemisphere is tilted towards the Sun **[1]** and winter is when a hemisphere is tilted away from the Sun **[1]**.

Page 109

1. a) iii and iv [2]
   b) iv [1]
2. a) As we get older we lose our ability to hear higher pitched/ high frequency sounds **[1]**. Hence teenagers can hear the noise, but the adults cannot **[1]**.
   b) The loudspeaker vibrates **[1]**, hitting air particles which gain energy **[1]**. They move in waves to the teenager's ear **[1]**. The eardrum vibrates at the same frequency as the original sound **[1]**.
3. a) Light rays leave the object, pass through the pinhole and land on the screen. [1]
   b) Light rays cross over in the pinhole, so that ones from the top of the object are at the bottom of the image and vice versa. [1]
   c) i) Larger **[1]**   ii) Fainter **[1]**

Page 110

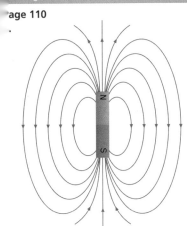

2. a) i; iv; vi [3]
   b) the electromagnet can be turned on and off [1]
3. A2 and A3 = 2 A, A4 = 4 A [2]
4. a) P = I x V so I = P/V =
      60 W / 230 V = 0.26 A [1]
   b) 60 x 200 = 12,000 Wh = 12 kWh [3]

Page 111

1.

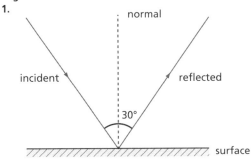

[3]

2. a)

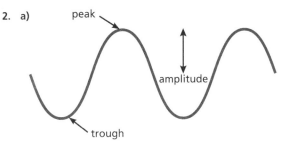

[3]

   b) 1 Hz [2]
3. red, orange, yellow, green, blue, indigo, violet (top to bottom on diagram) [4]

Page 112
1. c) cobalt [1]
2. C [1]
3. a) it will turn on/activate and attract the armature [1]
   b) it will be attracted to the electromagnet [1] and complete the second circuit [1]

Page 113
1. In a mirror, all the waves are reflected at the same angle [1]. This is specular reflection [1]. With paper, the rays hit the uneven surface and reflect in different directions. This is called scattered diffusion [1].
2. a) The sonar waves are transmitted from the boat, are reflected by the fish [1] and travel back to the boat. [1]
   b) Sonar waves sent down from the boat will be reflected back from the seabed. [1] Knowing the speed of the waves and measuring the time between transmission and reception will enable the distance to be calculated. [1]
   c) speed = distance/time so distance = speed × time. [1] Speed = 1500m/s and time = 0.5s to travel to sea bed and back. [1] so time to travel one way is 0.25s. [1] Therefore distance = 1500 × 0.25 = 375m. [1]
3. a) [3]

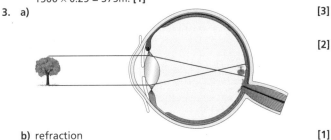

[2]

   b) refraction [1]

Page 114
1. a) i) B should be drawn inside the round bottomed flask or by the heating arrow. [1]
      ii) The distillate is the liquid collected at the end of the condenser. The label X should appear in the liquid, or be clearly labelled with a line to the collected liquid. [1]
   b) The cold water enters at the bottom of the condenser [1]. This is so that the cold water completely fills the condenser, which will make it more effective at cooling the gas. [1]
   c) 9 [1]
   d) Point Y is a liquid. At least 11 particles need to be drawn with at least 50% of the particles touching; Point Z is a gas. A minimum of five particles need to be drawn with no particles touching [**both diagrams need to be drawn correctly for 1 mark**].
2. a) 20 N [1]
   b) Michael is already using 20 N of force.
      Kerrie weighs 600 N, so 600 N / 2 = 300 N.
      Therefore 320 N of force used. [2]
      If part (a) is incorrect and in part (b) only uses 600 N, follow through and accept 300 N (with working).
      If part (a) is correct but in part (b) only uses 600 N, award 1 mark for working.
   c) i) pressure = force / area
         Force = 650 N. The area of 1 stilt is 9 cm². There are 2 stilts. Therefore total area = 2 x 9 cm² = 0.0018 m².
         650 N / 0.0018 = 361,111.1 N/m²
         [**1 mark for correct working, 1 mark for correct answer**]
         If working shows only 9 cm² used, award maximum 1 mark.
      ii) To exert less pressure she needs to use stilts with a larger surface area or reduce her weight. [1]
3. a) Breathing in is shown by the rubber sheet being pulled down [1]; volume in the bottle increases [1] and pressure decreases so air enters the balloons [1]. Breathing out is shown by the rubber sheet going up [1], volume decreases [1], pressure increases and forces air out of the balloons [1].
   b) i) asthma [1]
      ii) smoker [1]
      iii) normal [1]
   c) i) 10→8; 10→6; 8→6 [3]

Diffusion always moves from high to low concentration.

   ii) Glucose moves by diffusion **[1]** from a high concentration **[1]** to a low concentration **[1]**.

d) i) anaerobic makes lactic acid or aerobic doesn't make lactic acid **[1]**;
aerobic releases more energy or anaerobic releases less energy   **[1]**

   ii) humans produce lactic acid **[1]**; yeast produces alcohol **[1]**

Remember *aerobic* uses oxygen in *air*.

4. a) A new material has been deposited on the surface of the nail.   **[1]**

b) iron + oxygen + water **[1]** → (hydrated) iron oxide **[1]**

c) An iron nail will only rust if water and oxygen are both present / an iron nail will not rust if either water or oxygen are missing.   **[2]**

d) i) Naveen would see a reddish–brown metal forming on the nail **[1]**. She would also see the blue colour disappearing / a green solution forming **[1]**. [**Accept the colour of the nail has changed**]

   ii) iron displaced the copper from its compound (copper sulfate) and takes its place **[1]**; iron is higher in the reactivity series than copper **[1]**

5. a) Numbers of blue tits reduce **[1]** because fewer insects would be available to eat **[1]**. Numbers of moths increase **[1]** because there is less competition for their food from the leaf-eating insects **[1]**.

This question is about interdependence, i.e. how one organism in a food web affects another organism.

b)

	✓ or ✗
Respiration stores energy as light	✗
Photosynthesis uses energy from light	✓
Respiration breaks down large molecules to smaller molecules	✓
Photosynthesis creates organic molecules from inorganic molecules	✓
Respiration releases energy from chlorophyll	✗
Photosynthesis stores energy as light	✗

   [6]

c) i) extinction means all members of a species have died out worldwide   **[1]**

Remember, once an organism is extinct it cannot return.

   ii) a change in their environment **[1]** which happened rapidly **[1]**

d)

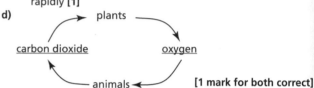

   [1 mark for both correct]

e) Most scientific discoveries are the result of a scientist building on the work of a previous scientist **[1]**. Scientists have their results checked by other scientists **[1]**. Scientists use data from experiments to check their ideas **[1]**.

6. a) i) North on the left end and South on the right.   **[1]**

   ii) All the carriages will now attract the ones next to them **[1]** because opposite poles attract. **[1]**.

b) A, steel; B, magnet; C, aluminium   **[2 marks for all 3 correct, 1 mark for 2 correct, no marks for only 1 correct]**.

c) cobalt and iron should be ticked   **[1]**

Glossary

a

absorb (*in physics*) waves of light taken in by a material and not reflected

abundance how available a substance is

abuse misuse of a drug

acid a compound containing hydrogen that can be replaced by a metal to form a salt

activation energy the amount of energy that has to be available before a chemical reaction can take place

addiction the need to keep taking a drug

aerobic respiration that uses oxygen

air pressure the force due to gases acting on an area

alkali soluble form of a base

alveoli small air sacs in the lungs

amplitude the maximum displacement of a wave from zero

anaemia deficiency disease caused by a lack of iron

anaerobic respiration that does not use oxygen

antagonistic pair two muscles that move a joint in different directions

anther male part of a flower that produces pollen

antibiotic a drug that kills some types of bacteria

antibody chemical produced by our body to destroy a microorganism

anus exit of the digestive system

asthma a condition that makes breathing difficult

atom the smallest component of an element that retains the element's chemical properties

attraction (*in physics*) the electric or magnetic force that acts between bodies with an opposite charge

b

bacteria a type of microorganism

base a solid chemical that reacts with an acid to form a salt and water

bioaccumulation build-up of toxic materials in a food web

biodiversity the range of different species in an ecosystem or the range of variation in a species

Brownian motion the random movement of particles in a fluid (liquid or gas) due to collisions with the atoms or molecules in the fluid

c

carbohydrate foods such as starch and sugar

carbon cycle a sequence where carbon is exchanged among the different spheres of the Earth (the atmosphere, lithosphere, hydrosphere and biosphere)

cartilage tissue that cushions the ends of bones at a joint

catalyst a chemical that lowers the activation energy of a reaction without being used up in the process

cell wall the supporting wall around a plant cell

ceramic a non-metallic solid that has been altered through heating

chemical potential energy the amount of energy stored in chemical bonds. This is released during combustion or respiration.

chemical reaction representation of a chemical reaction using the chemical formulae of the reactants and products

chlorophyll green chemical found in plants that aids photosynthesis

chloroplast place where chlorophyll is stored

chromosome a long strand of DNA

combustion the chemical reaction between fuel and oxygen

composite two or more substances with different properties that, when combined, form a new material with different properties

compression reduction in volume as a result of a pushing force

concentration the abundance of a substance in a given volume

condense to turn from a gas to a liquid

conductor has a low resistance to the flow of electric current or heat energy

consensus broad agreement

continuous variation variation that occurs across a range

consumers animals which consume food for energy

convex a lens which is thickest in the middle

cytoplasm the place where chemical reactions take place in a cell

d

DC motor a motor that can be run on direct current

density the amount of mass per unit volume

diaphragm domed shaped muscle used during breathing

diffusion (*in chemistry*) the process whereby molecules in a liquid or gas mix as a result of their random motion; (*in biology*) movement of molecules from a high to a low concentration

discontinuous variation variation that occurs in distinct groups

dispersal spreading of seeds over a wide area

displace the action of one metal replacing another in a compound

distillate the collected liquid resulting from evaporation and then condensation in distillation apparatus

DNA chemical that codes for life

drug a substance that changes the body

ductile capable of being drawn into wires

e

echo a reflected sound wave arriving back some time after the sound was originally made

ecosystem a system of interacting organisms that live in one area

egg cell female cell used for reproduction

electric field generated by electrically charged particles in a varying magnetic field

electromagnet a temporary magnet created by an electric current running through a wire

element a substance made up of only one type of atom, having the same number of protons in each nucleus

embryo a group of cells that will grow into a baby

environment the surroundings in a habitat

equilibrium where both potential states of a system are equal

equinox the day when day and night are the same length

evaporation when a liquid turns into a gas

extinction when all the members of a species are dead

f

fat a type of food that is used to store energy

fermentation the process of anaerobic respiration in yeast

fertile able to reproduce using sexual reproduction

fertilisation the joining together of a male and female sex cell

fibre indigestible material found in food

field lines demonstrate where a magnetic field is present

filtrate the liquid which passes through filter paper

filtration a method of separating an insoluble solid from a liquid

food chain picture showing which organisms eat each other in a habitat

food web many food chains linked together

force a push or pull resulting from the interaction between two objects

fossil fuels high energy substances made from the remains of plants and animals that died millions of years ago that have been subjected to immense pressure and high temperatures

frequency number of waves per second. Measured in Hz

fungi a type of microorganism

g

gene a single instruction found on a chromosome

gestation time period from fertilisation to birth

glucose food produced by photosynthesis

gravitational field strength the strength of an object's gravitational pull on objects

gravitational potential energy the amount of energy an object has due to its position in a gravitational field. The higher the object, the more energy

group (*in chemistry*) a column of elements in the periodic table

h

heredity how characteristics are passed from one generation to the next

i

igneous rock formed when magma cools and solidifies

immune system system in our bodies that protects us from invading microorganisms

inheritance how characteristics are passed from one generation to the next

insulator does not conduct heat and/or electric current

interdependent organisms which depend upon one another for their survival

interspecific variation variation between species

interference where waves are out of sync they can either cancel or enhance each other

intestine the tube that goes from the stomach to the anus

intraspecific variation variation within a species

j

joint where two bones are held together to allow movement

joule the unit of energy

k

kinetic energy the energy possessed by an object due to its motion

kwashiorkor a protein deficiency disease

l

lactic acid waste substance produced by anaerobic respiration in humans

ligament tissue that holds two bones together at a joint

light year an astronomical unit of distance representing the distance light travels in an Earth year

longitudinal (*in physics*) disturbances that move in the direction of travel

m

magnetic field the area around a magnet where it can affect other magnetic materials

malleable capable of being hammered into shape without cracking

medicine a drug used to treat or prevent disease

membrane controls what goes into and what leaves a cell

memory cell a type of white blood cell produced after vaccination that produces antibodies to destroy that type of invading microorganism

menstruation monthly loss of tissue from the lining of a uterus

metamorphic rock that is transformed by high temperature and pressure

minerals chemicals needed by our body

mitochondria place where energy is released inside a cell (respiration takes place)

moment a force multiplied by the distance to a pivot

monomer a building block of a polymer

n

nectar sweet substance in flowers that attracts insects

neutral a chemical that is neither an acid nor a base

neutralisation the process by which an acid is neutralised by a base leading to a solution with pH 7

Newton the unit of force

niches small areas within an ecosystem

nucleus part of a cell that contains the DNA

o

observer (*in physics*) a person making an observation from a particular frame of reference

oesophagus a tube that connects the mouth to the stomach

organ a group of different tissues working together

organ system a group of organs working together

ovary   place where female sex cells are made

oxidation   the gain of oxygen in a chemical reaction

oxygen   gas given out by green plants

p

pancreas   organ that produces digestive enzymes

parallel   in a circuit this means that the circuit branches

pathogen   disease-causing organism

peak   (*in physics*) the top of the wave

period   a horizontal row on the periodic table

photosynthesis   the process by which plants make food (glucose)

pivot   the point of rotation of a lever

plimsoll line   drawn on a ship to indicate how deep the hull of the ship is submerged

pollination   transfer of pollen from one flower to another

polymer   a substance made of many monomers chemically joined together

pressure   the ratio of force to the area or volume over which the force is applied

producers   plants which produce food by photosynthesis

products   new chemicals formed as a result of a chemical reaction

protein   food used for growing new cells

pyramid of numbers   the number of each organism in a food chain shown as a horizontal bar chart. The wider the bar, the more organisms there are

r

reactants   the chemicals that react together in a chemical reaction

reaction force   the force exerted on an object countering the gravitational force of that object

reactivity   a measure of how easily a chemical reacts with other chemicals

recreational drug   a drug used for pleasure

rectum   last part of the digestive system where water is absorbed

recycling   collecting, extracting and then using for another purpose to avoid waste of useful materials

reduction   the loss of oxygen from a molecule

reflection   the bouncing of a wave off a surface

refraction   the apparent bending of light as it moves between mediums with different optical densities

relative   (*in physics*) the description given to where an observer is in relation to a scenario, e.g. watching two cars passing on the road or being in one of the cars

repulsion   the electric or magnetic force that acts between bodies with the same charge

resistance   the opposition of a component to the electric current running through it, leading to the transfer of energy

respiration   breaking down food to release energy

rust   the product of the reaction between iron, oxygen and water

s

salt   the name given to the product formed by swapping the hydrogen in an acid with a metal

scurvy   a disease caused by a lack of vitamin C

sedimentary   rock type formed by the settling of layers of material followed by extreme pressure

series   a circuit which is in a single loop, with no branches

side effect   an unwanted effect of a drug on the body

solstice   the longest or shortest day

solvent   a liquid into which solutes can dissolve

species   a group of organisms that reproduce to produce fertile offspring

speed   how far an object travels divided by the time travelled

sperm   male cell used for reproduction

stigma   female part of a flower where pollen is deposited toma   small hole on the underside of a leaf

tomach   a bag-shaped organ that helps digest food

tomata   lots of small holes on the underside of leaves

sublimation   transformation from a solid to a gas, missing out the liquid state

t

tendon   tissue that attaches muscle to bone

testes   place where male sex cells are made

thermal decomposition   breakdown of a compound into products when heated

tissue   a group of the same type of cells

toxin   substance poisonous to humans

transverse   moving side to side at right angles to the direction of travel. Transverse waves look like a repeating sideways S

u

ultrasound   sound with a frequency of over 20,000 Hz

unicellular   organism that consists of just one cell

upthrust   a specific type of reaction force, the counterforce of water to the weight of an object in water

uterus   place where a baby grows inside a woman

v

vaccination   a means of making the body produce antibodies to destroy an invading microorganism

vacuole   storage space inside a plant cell

variation   differences between different organisms

vibration   the movement of an object back and forth from an equilibrium position

virus   a type of microorganism

vitamins   essential nutrients needed in small amounts by the body to work properly

w

weight   the force of a mass in a gravitational field

word equation   representation of a chemical equation using the names of the reactants and products

work   the transfer of energy as an object moves in the direction of the force

Index

Collins

KS3
Science

Workbook

Heidi Foxford, Emma Poole and Ed Walsh

Revision Tips

Rethink Revision

Have you ever taken part in a quiz and thought '*I know this*!', but no matter how hard you scrabbled around in your brain you just couldn't come up with the answer?

It's very frustrating when this happens, but in a fun situation it doesn't really matter. However, in tests and assessments, it is essential that you can recall the relevant information when you need to.

Most students think that revision is about making sure you **know** *stuff*, but it is also about being confident that you can **retain** that *stuff* over time and **recall** it when needed.

Revision that Really Works

Experts have found that there are two techniques that help with *all* of these things and consistently produce better results in tests and exams compared to other revision techniques.

Applying these techniques to your KS3 revision will ensure you get better results in tests and assessments and will have all the relevant knowledge at your fingertips when you start studying for your GCSEs.

It really isn't rocket science either – you simply need to:

- **test yourself** on each topic as many times as possible
- **leave a gap** between the test sessions.

It is most effective if you leave a good period of time between the test sessions, e.g. between a week and a month. The idea is that just as you start to forget the information, you force yourself to recall it again, keeping it fresh in your mind.

Three Essential Revision Tips

1. **Use Your Time Wisely**
   - Allow yourself plenty of time
   - Try to start revising six months before tests and assessments – it's more effective and less stressful
   - Your revision time is precious so use it wisely – using the techniques described on this page will ensure you revise effectively and efficiently and get the best results
   - Don't waste time re-reading the same information over and over again – it's time-consuming and not effective!

2. **Make a Plan**
   - Identify all the topics you need to revise (this Complete Revision & Practice book will help you)
   - Plan at least five sessions for each topic
   - A one-hour session should be ample to test yourself on the key ideas for a topic
   - Spread out the practice sessions for each topic – the optimum time to leave between each session is about one month but, if this isn't possible, just make the gaps as big as realistically possible.

3. **Test Yourself**
   - Methods for testing yourself include: quizzes, practice questions, flashcards, past-papers, explaining a topic to someone else, etc.
   - This Complete Revision & Practice book gives you seven practice test opportunities per topic
   - Don't worry if you get an answer wrong – provided you check what the right answer is, you are more likely to get the same or similar questions right in future!

Visit our website to download your free flashcards, for more information about the benefits of these revision techniques and for further guidance on how to plan ahead and make them work for you.

collins.co.uk/collinsks3revision

Contents

Cells – the Building Blocks of Life

1 The diagram shows a cell.

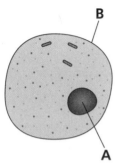

B

A

a) Name the structures **A** and **B**.

A: .... B: .... [2]

b) Describe the function of the part of the cell labelled B.

.... [1]

2 The table below gives information about a plant cell's structure and function. Complete the table by filling in the missing information.

Name of Structure	Function
Cell wall	
	Place where lots of chemical reactions take place.
Chloroplasts	
	Contains cell sap and keeps the cell firm.

[4]

3 The table below shows the steps that need to be followed to view a cell using a light microscope. Number the steps **1** to **5** to put them in the correct order.

Description of Step	Order (1–5) Starting with 1
Place the slide on the stage of the microscope.	
Stain the sample.	
Illuminate and focus the microscope.	
Place the cell sample on a slide.	
Cover the sample with a coverslip.	

[5]

4 The diagram below shows a cell.

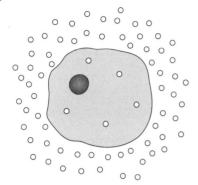

Key
○ glucose molecule

a) Glucose will diffuse into the cell in the diagram. Explain why glucose diffuses into the cell using information from the diagram.

....

.... [1]

b) The cell shown is an animal cell, not a plant cell. Suggest how we can tell it is an animal cell not a plant cell.

.... [2]

5 a) Flowering plants reproduce sexually through a process called pollination. Describe what happens in pollination.

....

.... [2]

b) Campanula flowers contain nectar. Explain how this helps increase the chances of pollination.

....

.... [2]

6 Sexual reproduction in humans involves sperm and egg cells. Name the following:

Male reproductive organ that produces sperm cells: ....

Female reproductive organ that produces egg cells: .... [2]

7 List the following words in the correct order in which they happen in human reproduction.

gestation   fertilisation   birth

.... [3]

Total Marks .... / 24

Biology

Eating, Drinking and Breathing

1 The diagram shows parts of the breathing system.

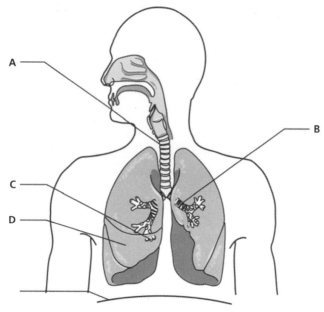

a) Name the parts of the breathing system labelled **A–D**.

A: .... B: .... C: .... D: .... [4]

b) Give the function of the part labelled **C**.

.... [1]

2 Complete the sentences using the words below.

increases   decreases   outwards   inwards   contracts   relaxes

When we inhale, the ribcage moves upwards and .... The diaphragm

.... and flattens. This .... the volume of the chest cavity

and .... the pressure, causing air to move into the lungs from outside. [4]

3 Describe the effect of asthma on the breathing system.

.... [2]

4 Chemicals in tobacco smoke damage the cilia in the breathing tubes. Explain how damage to the cilia causes the smoker to develop a cough.

....

....

.... [2]

5 Draw lines to match each food group to its correct use in the body.

Food Group		Use in Body
Carbohydrates		Important for growth and repair
Proteins		Provide energy
Fats		Helps undigested food pass quickly through the gut
Fibre		Provide a reserve energy supply and insulation

[4]

6 The following words are all parts of the digestive system. Put them in the order in which food passes through them. The first one has been done for you.

anus   large intestine   mouth   oesophagus   rectum   small intestine   stomach

mouth

[5]

7 A student has been diagnosed with a nutritional deficiency called anaemia. The student's doctor has advised the student to eat more nuts, beans, green vegetables and red meat.

Suggest why.

[2]

8 The digestive system produces biological catalysts called digestive enzymes.

a) Explain the function of digestive enzymes.

[2]

b) What are enzymes made from? Tick the correct box.

Carbohydrate ☐   Fat ☐

Protein ☐   Keratine ☐   [1]

Total Marks _____ / 27

Biology

Getting the Energy your Body Needs

1 The table below shows some statements about respiration.

Put a tick in the correct column to say whether each statement is true or false.

Statement	True	False
Respiration is a chemical reaction.		
Respiration takes place in the cells of plants and animals.		
Respiration only takes place in the lungs.		
Respiration releases energy.		

[4]

2 Complete the word equation for aerobic respiration.

Glucose + _____ ⟶ Water + _____ [2]

3 Humans are able to respire aerobically and anaerobically.

a) Write the word equation for anaerobic respiration in humans.

_____ [2]

b) Compare aerobic respiration with anaerobic respiration in humans.

_____

_____

_____ [3]

4 Which of the following are functions of the skeleton? Tick four answers.

To provide support ☐

To protect organs ☐

To make red blood cells ☐

To insulate the body ☐

To make enzymes ☐

To allow the body to move ☐ [4]

5 The diagram below shows the quadriceps and hamstrings muscles in a human leg. They are antagonistic muscles.

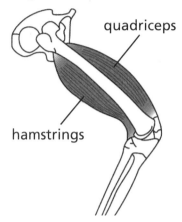

quadriceps

hamstrings

a) Complete the table by writing 'contracts' or 'relaxes' to show the action of each muscle when bending and straightening the leg.

Muscle	Bending the Leg	Straightening the Leg
Quadriceps		
Hamstrings		

[4]

b) The end of the large bones in the leg are covered in cartilage.
Explain why cartilage is important.

....

.... [1]

c) The knee joint contains tendons and ligaments.
Explain the difference between a tendon and a ligament.

....

.... [2]

Total Marks .... / 22

Biology

Looking at Plants and Ecosystems

1 **a)** Complete the sentences using the words below.

chloroplast   **heat**   **protein**   **light**   **glucose**   **minerals**

Plants use energy carried by .... from the sun. Plants use this energy

to make ....   [2]

b) Complete the word equation for photosynthesis.

$$\text{.........................} + \text{Water} \xrightarrow[\text{chlorophyll}]{\text{light}} \text{.........................} + \text{Oxygen}$$ [2]

2 Complete the table to show how leaves are adapted for photosynthesis. The first one has been done for you.

Adaptation of Leaf	Function
Are flat and broad	To *give a large surface area for absorbing light.*
Lower surface contains stomata	
Contain xylem and phloem tubes	

[2]

3 Leaves have small holes called stomata.

a) Describe the function of stomata.

.... [1]

b) Explain why stomata help the plant survive.

....

.... [2]

4 The diagram shows a food web in a pond ecosystem.

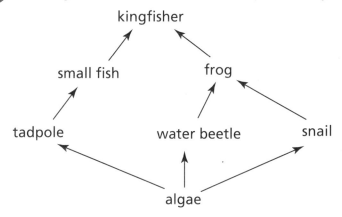

a) Draw **two** possible food chains using the information given in the food web.

[2]

b) Mercury is found in the pond. High concentrations of mercury are toxic to most organisms. Explain how the kingfisher could end up poisoned by mercury.

[4]

5 Plants and animals depend on each other for survival. What scientific word best describes this relationship? Tick one box.

Interdependency

Dependency

Biodiversity

Sustainability

[1]

Total Marks _____ / 16

Biology

Variation for Survival

1. Complete the table to show whether each statement about genes and inheritance is true or false.

Put a tick in the correct column for each statement.

Statement	True	False
A baby inherits half its genetic information from each parent.		
Brothers and sisters inherit the same genetic information from their parents.		
Genetic information is stored within chromosomes, found within the nucleus of cells.		
DNA is only found in egg and sperm cells.		

[4]

2. The diagram shows the genetic material found inside the nucleus of a cell.

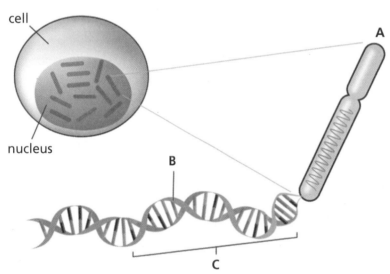

cell

nucleus

A

B

C

Name the structures labelled **A**, **B** and **C**.

A: ....

B: ....

C: ....

[3]

3. Human body cells contain 46 chromosomes. Human sperm and egg cells only have 23 chromosomes. Explain why.

....

....

[2]

4 Draw one line from each characteristic to the correct type of variation.

Characteristic	Type of Variation

Foot length

Blood group

Hamster weight

Fur colour

Discontinuous

Continuous

[4]

5 Explain why variation within a species is very important when the environment changes.

_____

_____

_____ [2]

6 A gene bank is a place where scientists store seeds and cells. Suggest two advantages of having seed banks.

_____

_____ [2]

7 Different scientists were involved in the discovery of DNA. Draw lines to match each scientist with their part in the discovery.

Scientist	Discovery

Watson and Crick

Maurice Wilkins

Rosalind Franklin

Made X-ray images of DNA

Produced evidence to support Watson and Crick's theory

Developed a theory for the structure of DNA

[3]

Total Marks _____ / 20

Biology

Our Health and the Effects of Drugs

1 Which of the following statements about drugs are true and which are false? Put a tick in the correct column for each statement.

Statement	True	False
A drug is a chemical that alters the way your body or mind works.		
Only a few drugs cause side effects.		
Caffeine, alcohol and nicotine are all drugs.		
Only illegal drugs are additive.		

[4]

2 Draw lines to match the type of drug to the effect it has on the body.

Type of Drug	Effect on the Body
Stimulant	Makes a person feel relaxed and drowsy.
Depressant	Reduces pain and inflammation.
Painkiller	Makes a person feel energetic and alert.
Hallucinogen	Makes a person hear or see things that are not real.

[4]

3 Give two **long term** effects of drinking too much alcohol.

....

....

[2]

4 Name the three main types of microbes.

....

[3]

5 Complete the table to show how each part of the body helps protect the body from disease. The first one has been done for you.

Part of Body	Defence
Eyes	Produce tears that contain a chemical to kill microbes.
Ears	
Nose and throat	

[2]

6 White blood cells protect us against infection from microbes.

Describe two ways in which white blood cells help fight infection from microbes.

_____

_____

[2]

7 During the Covid-19 pandemic, a number of pharmaceutical companies produced vaccines.

a) Explain how a vaccination works.

_____

_____

_____

_____

[2]

b) Antibiotics are not given to people infected with the Covid-19 virus.

Explain why.

_____

_____

[2]

Total Marks _____ / 21

Mixing, Dissolving and Separating

1 Identify the impure substance from the list of substances below. Tick one box.

Oxygen gas ☐

Ice ☐

Orange juice ☐

Sodium chloride salt ☐ [1]

2 A student investigates the inks found in five felt-tip pens. The results of the experiment are shown in the diagram below.

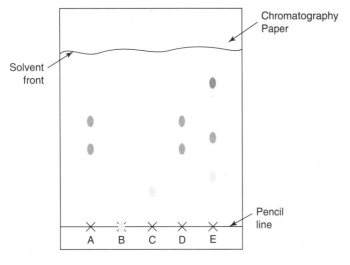

a) Which ink is pure?

.... [1]

b) Which inks are identical to each other?

.... and .... [1]

c) Why does ink B not move during the experiment?

.... [1]

d) Which ink contains the most colours? Explain your answer.

....

.... [2]

3 Filtration can be used to separate insoluble solids from liquids.

Use the labels in the box to label the parts **A–D** shown in the diagram.

Filter paper	**Filter funnel**	**Filtrate**
Residue	**Measuring cylinder**	

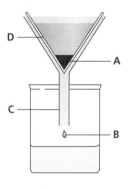

A: .... B: ....

C: .... D: .... [4]

4 When substances burn, they react with oxygen in the air.

A teacher reacts 1.2g of magnesium with exactly 0.8g of oxygen.

a) Name the compound formed in this reaction.

.... [1]

b) Deduce the mass of the compound formed in this reaction.

.... [1]

5 Methane, CH_4, is the fuel used in Bunsen burners. Identify the products of the reaction between methane and a good supply of oxygen. Tick one box.

Carbon monoxide and water ☐

Carbon and hydrogen ☐

Carbon dioxide and water ☐

Methane and carbon dioxide ☐ [1]

Total Marks .... / 13

Chemistry

Elements, Compounds and Reactions

1 Draw lines to link each key word to its definition.

Key Word	Definition
Atom	Contains different types of atoms joined together by chemical bonds
Element	A very small particle
Compound	A column in the periodic table
Group	Contains only one type of atom

[4]

2 Glucose has the formula $C_6H_{12}O_6$.

a) How many hydrogen atoms are present in one molecule of glucose?

.... [1]

b) How many different elements are present in glucose?

.... [1]

c) How many atoms are present in total in one molecule of glucose?

.... [1]

3 Elements can be represented by one- or two-letter codes called symbols. Draw lines to match each element's name to its symbol.

Name	Symbol
Copper	Ag
Silver	S
Magnesium	Cu
Sulfur	Na
Sodium	Mg

[5]

4 A student places a piece of magnesium into a beaker of dilute hydrochloric acid. The student observes that the magnesium slowly disappears and bubbles of hydrogen form. A solution of a salt is also made.

a) Identify the reactants in this experiment.

_____ [2]

b) Name the salt made in this reaction.

_____ [1]

c) Explain why the student observes bubbles being made in this reaction.

_____ [1]

d) Magnesium is a metal. Tick the boxes to show the properties magnesium will have.

Conducts electricity well ☐

Shiny when cut ☐

Poor thermal conductor ☐

Gas at room temperature ☐

Low melting point ☐

High boiling point ☐ [3]

5 A student heats a small piece of copper metal in a Bunsen burner flame for several minutes. The metal changes colour from brown to black.

a) Identify the name of the compound formed in this reaction. Tick one box.

Rust ☐   Copper oxygen ☐

Copper oxide ☐   Copper sulfate ☐ [1]

b) Identify what has happened to the copper in this reaction. Tick one box.

Neutralisation ☐   Oxidation ☐

Reduction ☐   Thermal decomposition ☐ [1]

Total Marks _____ / 21

Chemistry

Explaining Physical Changes

1 Water can exist in each of the three states of matter.

a) In which state or states:

  i) are the particles close together?

  .... [2]

  ii) do the particles have the most energy?

  .... [1]

  iii) does the substance have a fixed shape and volume?

  .... [1]

  iv) does the substance have a fixed volume but not a fixed shape?

  .... [1]

  v) does the substance have the lowest density?

  .... [1]

b) Name the change of state that happens when solid carbon dioxide is heated and turns straight into gaseous carbon dioxide.

.... [1]

2 A student stands in front of a window and notices that the small dust particles in the air are moving around randomly. Explain why the dust particles are moving around.

....

....

.... [2]

3 A student walks past the school canteen bins on a summer's day and notices the smell of the rubbish in the bins. Explain how the student can smell the rubbish in the bins.

....

.... [2]

4 A student carries out a series of experiments to find out if the type of metal added to a beaker of dilute hydrochloric acid affects the temperature change of the reaction.

The results of her investigation are shown below.

Metal	Temperature at the Start (°C)	Temperature at the End (°C)	Temperature Change (°C)
Magnesium	21.0	27.5	+ 6.5
Copper	21.0	21.0	0.0
Zinc	20.5	25.5	
Iron	21.0	23.0	+ 2.0

a) Identify the independent variable in this investigation.

_____ [1]

b) Identify the dependent variable in this investigation.

_____ [1]

c) Explain why there is no temperature change for the copper experiment.

_____ [1]

d) Complete the table by calculating the temperature rise for the reaction between zinc and hydrochloric acid. [1]

5 A student places a green gel air freshener in the corner of a room.

The student notices that the green solid in the gel air freshener gets smaller over time. The solid gel turned into a gas that smelt nice.

Identify the type of change that has taken place. Tick one box.

Evaporation ☐

Melting ☐

Deposition ☐

Sublimation ☐ [1]

Total Marks _____ / 16

Chemistry

Explaining Chemical Changes

1 A teacher investigates a selection of everyday substances using Universal Indicator (U.I.) solution.

a) Complete the table to show the pH number, colour with Universal Indicator and type of solution for this selection of everyday substances.

Everyday Substances	pH Number	Colour with U.I.	Type of Solution
Lemon juice	2		
Hair dye	11		
Seawater		Green	
Lemonade		Orange	

[4]

b) Explain why the teacher wore goggles and gloves when she tested the hair dye.

i) She wore goggles to _____

ii) She wore gloves to _____ [2]

2 Bases react with acids to make salts.

a) Name the salt formed when sodium hydroxide reacts with sulfuric acid.

_____ [1]

b) Complete the equation to show the reaction between calcium carbonate and hydrochloric acid.

calcium carbonate + hydrochloric acid →

_____ + water + _____ [2]

3 Universal Indicator or blue litmus can be used to identify an acidic solution.

a) Name the colour blue litmus turns when it is placed into hydrochloric acid.

_____ [1]

b) Why is Universal Indicator more useful than a pure indicator like blue litmus?

_____ [1]

4 Identify which of the following is a property of a catalyst. Tick one box.

Increases the rate of reaction ☐

Decreases the rate of reaction ☐

Increases the temperature of the reaction ☐

Increases the time it takes for the reaction to happen ☐ [1]

5 Hydrogen reacts with oxygen to make water, H_2O.

a) Complete the equation below that sums up the reaction between hydrogen and oxygen by adding state symbols.

$2H_2$ .... $+ O_2$ .... $\rightarrow 2H_2O$ .... [3]

b) Describe what the state symbol (aq) indicates about a substance.

.... [2]

6 Ethane, C_2H_6, reacts with a good supply of oxygen to produce water and carbon dioxide as the only products.

Identify the balanced symbol equation for this reaction. Tick one box.

$C_2H_6 + 3\frac{1}{2}O_2 \rightarrow 2CO_2 + H_2O$ ☐

$C_2H_6 + 3O_2 \rightarrow 2CO_2 + 3H_2O$ ☐

$C_2H_6 + 3\frac{1}{2}O_2 \rightarrow 2CO_2 + 3H_2O$ ☐

$C_2H_6 + O_2 \rightarrow 2CO_2 + 3H_2O$ ☐ [1]

7 A teacher shows her class some powdered copper carbonate powder and a beaker of sulfuric acid.

a) Identify the pH of the sulfuric acid. Tick one box.

7 ☐   14 ☐   1 ☐   8 ☐ [1]

b) A student suggests that when the copper carbonate is added to the beaker, he will see bubbles. Is the student correct? Explain your answer.

....

.... [2]

Total Marks .... / 21

Chemistry

Obtaining Useful Materials

1 Draw lines to link each key word to its meaning.

Key Word	Meaning
Composite	Small molecules
Polymer	When a more reactive metal takes the place of a less reactive metal in its compound
Monomers	A material that is made from two or more different materials joined together
Displacement	A material made by chemically joining together many smaller molecules

[4]

2 The table below shows what happens when four different metals are added to a test tube containing dilute hydrochloric acid.

Metal	Observations
Iron	A few bubbles
Zinc	Many bubbles
Copper	No reaction
Magnesium	Lots of bubbles and the test tube becomes very warm

a) Place these metals in order of reactivity from the most reactive to the least reactive.

Most reactive _____

_____

_____

Least reactive _____ [1]

b) Write a word equation for the reaction between magnesium and hydrochloric acid.

_____ [2]

c) Explain why magnesium cannot be extracted from its ore, magnesium oxide, by heating it with carbon.

_____ [1]

3 Complete the word equation to show the displacement reaction between magnesium and copper sulfate.

magnesium + _____ → _____ + _____ [1]

4 A student carries out a number of experiments using different metal elements and solutions of their metal sulfates to find out whether a displacement reaction takes place or not. If a displacement reaction takes place, the student writes **yes** in the table. If there is no reaction, the student writes **no** in the table.

Each metal is represented by a letter. The letter is not the symbol of the element.

Metal/Metal Sulfate Solution	ASO_4	BSO_4	CSO_4	DSO_4
A	x	no	no	no
B	yes	x	no	no
C	yes	yes	x	yes
D	yes	yes	no	x

a) Use the results to place the metals into an order of reactivity from the most reactive to the least reactive.

Most reactive _____

_____

_____

Least reactive _____ [1]

b) Identify the letter that could be representing the element copper.

_____ [1]

Total Marks _____ / 11

Chemistry

Using our Earth Sustainably

1 There are three types of rock.

a) Draw lines to link each type of rock to its description.

Type of Rock	Description
Sedimentary	A very hard rock with interlocking crystals
Metamorphic	A hard rock which often has layers
Igneous	A soft rock which may contain fossils

[3]

b) Explain why the fossilised remains of plants are never found in igneous rocks.

_____

_____

[2]

c) Identify the conditions that would be best at turning the sedimentary rock, chalk, into the metamorphic rock, marble. Tick one box.

High pressure and low temperatures ☐

High pressure and high temperatures ☐

Low pressure and low temperatures ☐

Low pressures and high temperatures ☐

[1]

2 The Earth's crust and atmosphere have different compositions.

a) Name the most abundant element in the Earth's atmosphere.

_____ [1]

b) Name the most abundant element in the Earth's crust.

_____ [1]

3 The level of carbon dioxide in the atmosphere has increased from 280ppm to 415ppm since 1700.

a) Calculate the increase in the level of carbon dioxide since 1700.

_____ ppm [1]

b) Identify one reason why the level of carbon dioxide in the atmosphere has increased. Tick one box.

Plants carrying out photosynthesis ☐

Burning petrol in cars ☐

The Earth warming up ☐

The Moon moving round the Earth once every 28 days ☐ [1]

4 Aluminium cans can be made from aluminium that has been extracted from aluminium ores or by recycling old aluminium cans.

Explain why recycling aluminium cans is better for the environment than extracting the aluminium from its ores.

_____

_____ [2]

5 The Earth has a layered structure. Identify the outer layer of the Earth. Tick one box.

Inner core ☐   Outer core ☐

Mantle ☐   Crust ☐ [1]

6 The element carbon is common on Earth.

a) Explain how carbon is removed from the Earth's atmosphere.

_____

_____ [2]

b) Explain how living plants return carbon to the Earth's atmosphere.

_____

_____ [2]

Total Marks _____ / 17

Forces and their Effects

1 The diagram shows the forces acting on a car that is travelling along a level road in a straight line.

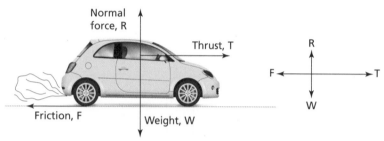

a) State two things that must be true about the weight, W, compared with the normal (or reaction) force, R.

....

.... [2]

b) If the car is travelling in a straight line and at a constant speed, what must be true about the thrust, T, compared with the friction, F?

.... [1]

c) If the car is accelerating forwards, what must be true about the thrust, T, compared with the friction, F?

.... [1]

d) Suggest, with reference to the forces shown on the diagram, why car designers try to make cars as streamlined as possible.

....

.... [1]

2 This pole is being used to move a boulder. Pushing down on the left-hand end of the pole will apply a force upwards on the boulder.

a) Which part of this arrangement is the pivot?

.... [1]

b) Compare the size of the force exerted on the left-hand end of the pole with the size of the force the right-hand end of the pole applies to the boulder.

_____ [1]

c) Explain how the arrangement of the equipment shown could be altered so that the same force applied on the left-hand end of the pole could produce a larger force on the boulder.

_____

_____ [2]

3 This question is about using a lever to apply a turning force.

a) Write down the equation that links the distance to the fulcrum, the force applied and the turning effect of the force.

_____ [1]

b) If the force is measured in Newtons and the distance in metres, what units will the turning effect be measured in?

_____ [1]

c) If the force applied is 50N and the lever is 75cm long, what will the turning effect be?

_____ [1]

4 This seesaw is balanced. The weight of the left-hand crate, W_1, is 50N and the weight of the right-hand crate, W_2, is 200N. The distance, D_1, from the fulcrum to W_1 is 3.2m.

a) Calculate the moment due to crate W_1.

_____ [2]

b) If the seesaw is balanced, what must be true about the moment due to crate W_1 and the moment due to crate W_2?

_____ [2]

c) Calculate the distance, D_2, from the fulcrum to crate W_2.

_____ [2]

Total Marks _____ / 18

Physics

Exploring Contact and Non-Contact Forces

1 This picture shows a children's slide.

a) Identify a part of the slide where it is important that friction is as low as possible and explain why.

.... [2]

b) Identify a part of the slide where it is important that friction is as great as possible and explain why.

.... [2]

2 Tick the correct column for each force to say which are contact forces and which are non-contact forces.

Force Acting	Contact Force	Non-contact Force
A person pushing down on a door handle.		
A charged rod attracting bits of tissue paper.		
An apple falling towards the ground.		
A match being struck on a matchbox.		

[4]

3 Add words to complete these sentences.

Gravity is a .... which attracts masses towards each other. The

space in which .... acts on objects is called the gravitational

.... The strength of a gravitational .... is measured

in .... per kilogram (N/kg). [5]

4 The diagram shows a concrete slab. Its dimensions are 0.4m × 0.4m × 0.05m and it has a weight of 200N.

a) Calculate the area of the largest face of the slab.

.... [2]

b) Calculate the pressure applied to the ground if the slab was placed with the largest face against the ground.

.... [2]

c) The slab is now turned so that it is standing with one of the narrow faces against the ground. Calculate the area of this face.

.... [2]

d) Calculate the pressure it exerts on the ground now.

.... [2]

e) Explain why the answer to part d) is much greater than the answer to part b).

....

.... [2]

5 When ice is placed in water it floats.

a) What does this show about the density of ice compared with the density of water?

.... [1]

b) Why is this unusual compared with the solid and liquid forms of most materials?

....

.... [1]

Total Marks .... / 25

Physics

Motion on Earth and in Space

1 Complete this table to show the missing values.

Object	Distance Travelled	Time Taken	Speed
Cyclist	100m		10m/s
Train		2 minutes	40m/s
Sound wave		10s	330m/s
Runner	400m		4m/s

[4]

2 These graphs show the motion of vehicles on two different journeys. Distance is measured in metres and time is measured in seconds.

A

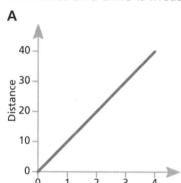

B

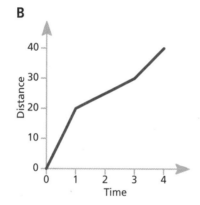

a) Which of these describes the motion of the car in Graph **A**? Tick the correct answer.

Stationary ☐   Moving at a steady speed ☐

Accelerating ☐   Decelerating ☐   [1]

b) Calculate the speed of the vehicle in Graph **A** in m/s.

.... [1]

c) Describe the motion of the vehicle in Graph **B**.

....

.... [1]

d) Calculate the speed of the vehicle during each stage of the journey shown in Graph **B** in m/s.

.... [3]

e) Calculate the average speed of the vehicle for the journey shown in Graph **B** in m/s.

.... [1]

3 This diagram shows the Earth tilted upon its axis.

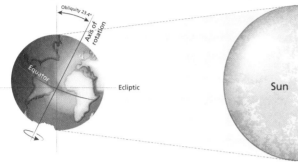

a) Explain how the tilt of the Earth on its axis causes variation in the seasons.

....

....

.... [2]

b) Looking at the arrangement shown, which season of the year is this for countries:

i) in the Northern Hemisphere? ....

ii) in the Southern Hemisphere? .... [2]

c) If the Earth was tilted on its axis by a lesser amount, suggest how this would alter seasonal variation on the Earth.

....

.... [2]

4 An astronaut wearing a space suit has a mass of 100kg. Use the following equation to answer the questions: **weight = mass × gravitational field strength**

a) Calculate the astronaut's weight on Earth, where the gravitational field strength is 10N/kg.

.... [2]

b) Calculate the astronaut's weight on the Moon, where the gravitational field strength is 1.7N/kg.

.... [2]

c) Calculate the astronaut's weight on the International Space Station, where the gravitational field strength is 8.9N/kg.

.... [2]

Total Marks .... / 23

Physics

Energy Transfers and Sound

1. Tick the correct column to indicate whether each of these fuels is renewable or non-renewable.

Fuel	Renewable	Non-renewable
Coal		
Wind		
Nuclear		
Natural gas		
Solar		
Oil		

[6]

2. This electric kettle is rated at 2kW.

   a) State this rating in watts.

   _____ [1]

   b) State how many joules of energy it transfers every second.

   _____ [1]

   c) Calculate the amount of energy it transfers in one minute.

   _____ [2]

   d) The appliance is designed to transfer energy into the water in the kettle. However, some of the energy ends up being transferred elsewhere. Explain why this is.

   _____ [1]

3. Some friends are going camping. They need to decide how to cook food and heat water. The options they have are to take a camping gas stove or to gather wood to light a fire.

   a) They know that the energy content of camping gas is 36kJ/g and that of wood is 12kJ/g. How many times greater is the energy content of the camping gas than that of wood?

   _____ [1]

   b) Suggest two reasons why they might select wood as the best fuel to use.

   _____ [2]

   c) Suggest two reasons why they might select camping gas as the best fuel to use.

   _____ [2]

4 Add words to complete the sentences.

Sounds are made when something vibrates, such as the _____ of a drum or the

_____ of a guitar. These _____ travel through the air because

of particles vibrating. Sound needs a medium to _____ through. We can hear

_____ because the vibrations transfer _____ to our ears.   [6]

5 A website states that the audible range of humans is from 20Hz to 20kHz.

a) Express 20kHz in Hz. _____   [1]

b) Explain what is meant by 'audible range'.

_____   [1]

c) Humans cannot hear ultrasonic sounds. Explain what this means about the frequencies of
ultrasonic sounds.

_____   [1]

6 Marcus is watching a cricket match but is standing quite a long way from the wicket. He notices
that when the ball is hit, he sees the bat and ball connect before he hears the sound.

a) Explain what this shows about the speed of sound compared with the speed of light.

_____   [1]

b) Suggest what he would notice about the gap between seeing the ball being hit and
hearing it if he was standing closer to the wicket.

_____   [1]

7 A student has a glass of water which is at room temperature. He adds a couple of ice cubes
to it and a couple of minutes later tastes the water. The ice cubes are smaller and the water
is cooler. Tick which of the following statements are true.

Energy has been transferred from the ice cubes to the water.   ☐

The internal energy of the water is less than it was before the ice was added.   ☐

Some of the ice has melted; this change required energy to be
transferred to the ice.   ☐   [2]

Total Marks _____ / 29

Magnetism and Electricity

1 Look at this circuit and write which of these statements are true and which are false.

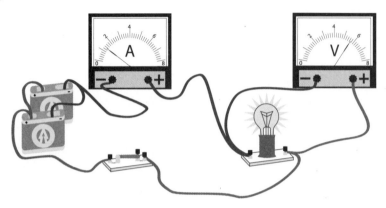

a) The ammeter is connected in series with the bulb. ....

b) The voltmeter is connected in series with the bulb. ....

c) The batteries are connected to each other in parallel. ....

d) Dividing the voltmeter reading by the ammeter reading will give a value for the resistance of the bulb. .... [5]

2 The diagram shows two bulbs in parallel. There are four ammeters in the circuit. Ammeter A_1 shows a reading of 0.2A and ammeter A_2 shows a reading of 0.1A.

a) State the reading on ammeter A_4 and explain your answer.

....

.... [2]

b) Calculate the reading on ammeter A_3 and explain your answer.

....

....

.... [3]

3 The diagram shows two bulbs in series. Also in the circuit are three voltmeters. V_1 shows a reading of 6V and V_2 shows a reading of 3.2V.

a) Calculate the reading on voltmeter V_3 and explain your answer.

....

.... [2]

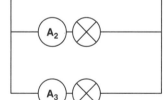

b) The potential differences across the two bulbs are not identical. Suggest what this shows about the bulbs.

_____ [2]

4 Explain how two bar magnets can be used to demonstrate magnetic attraction and repulsion.

_____

_____ [2]

5 A horseshoe magnet has a north pole and a south pole. Draw a diagram to show the magnetic field lines around a horseshoe magnet.

[2]

6 A student has been given a tray of materials and told that some of them are magnets, some are magnetic materials and some are non-magnetic. She has also been given a bar magnet. Describe how she could use the magnet to separate the objects into the three groups.

_____

_____

_____

_____ [3]

7 A student has made a simple electromagnet by coiling wire around a wooden rod and attaching the ends of the wire to a battery. The electromagnet works but isn't very strong. Describe three ways of modifying the electromagnet to make it stronger.

_____

_____

_____ [3]

Total Marks _____ / 24

Physics

Waves and Energy Transfer

1 Look at these three waves.

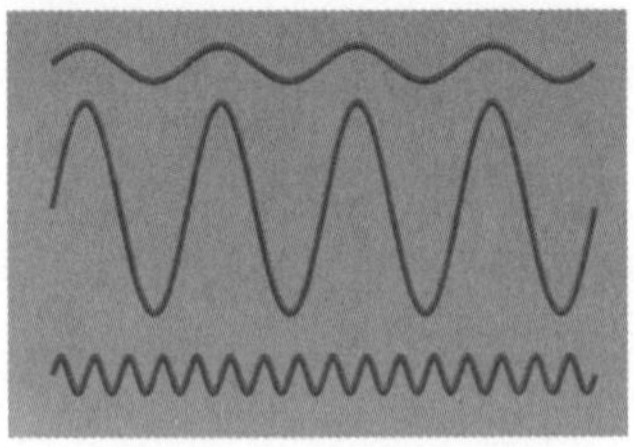

A

B

C

a) Which of the three waves has the greatest amplitude? _____ [1]

b) Which of the three waves has the shortest wavelength? _____ [1]

c) On wave **B**, mark:　**i)** a peak　　**ii)** a trough [2]

2 State suitable units to measure each of the following:

a) Frequency _____ **b)** Wavelength _____

c) Amplitude _____ **d)** Time _____ [4]

3 These diagrams show two different types of wave.

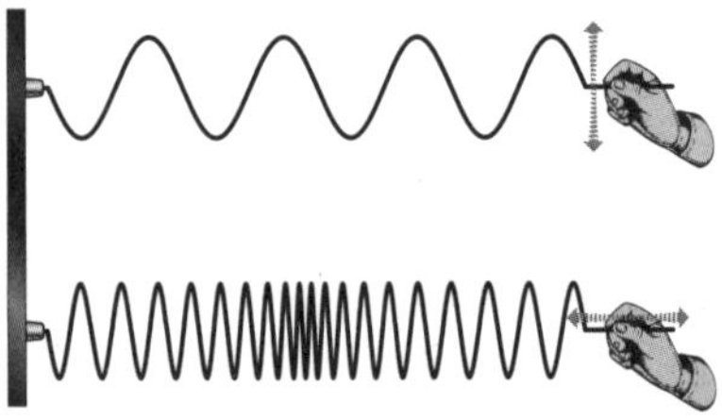

A

B

Next to each hand is an arrow which shows how the hand is moving to make the wave.

a) Identify which of these waves is a longitudinal wave and which is a transverse wave.

Longitudinal = _____　　Transverse = _____ [2]

b) For each wave, compare the direction in which energy is transferred to the wave with the direction in which the wave travels.

_____

_____

_____ [2]

4 The diagram shows a ray of light being reflected by a plane mirror.

a) Angles i and r are both measured from a dotted line. State the correct name for this line.

_____ [1]

b) Suggest what is true about the size of the angles marked i and r.

_____ [1]

c) A person looking into the mirror sees an image of the object at the place marked 'Image'. Explain why the ray from the image to the mirror is shown as a dotted line instead of a solid line.

_____

_____

_____

_____ [3]

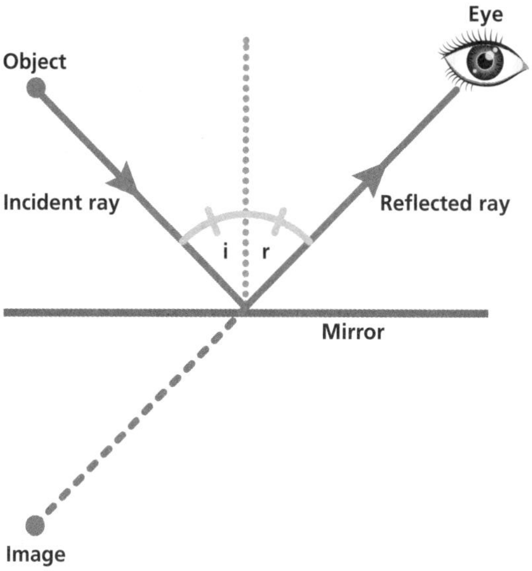

Object

Eye

Incident ray

Reflected ray

i   r

Mirror

Image

5 This diagram shows how light rays are refracted by a convex lens.

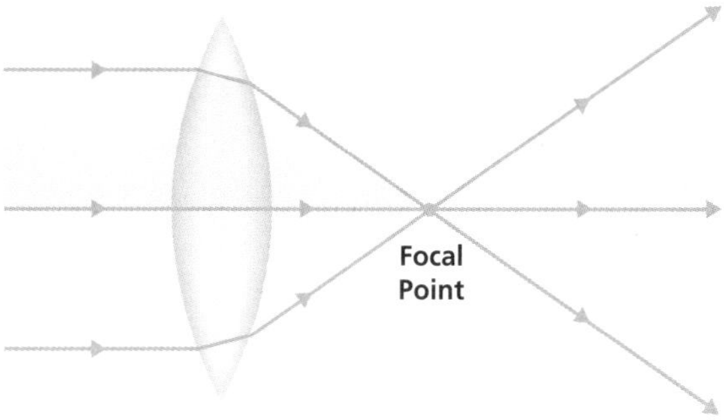

Focal
Point

a) What is the correct term to describe what is happening to light rays when they pass through the lens? Tick the correct answer.

Specular reflection ☐   Diffuse reflection ☐

Refraction ☐   Scattering ☐   [1]

b) Explain why this type of lens is also known as a converging lens.

_____ [1]

c) Suggest how this diagram would need to be altered if the convex lens was thinner than this one and less curved on the faces.

_____

_____ [2]

Total Marks _____ / 21

Mixed Test-Style Questions

1 A student adds an excess of powdered calcium carbonate, CaCO₃ to a beaker containing dilute hydrochloric acid. One of the products formed in the reaction is the salt calcium chloride.

a) Describe how the particles are arranged in the powdered calcium carbonate.

_____

☐ 1 mark

b) Explain why the student observed bubbles being made during this reaction.

_____

☐ 1 mark

c) Describe how the student could get a sample of pure dry calcium chloride from the reaction mixture.

_____

_____

_____

☐ 3 marks

TOTAL

☐

5

2 A teacher carries out a distillation experiment to separate a mixture of salt and water.

The diagram below shows the distillation equipment.

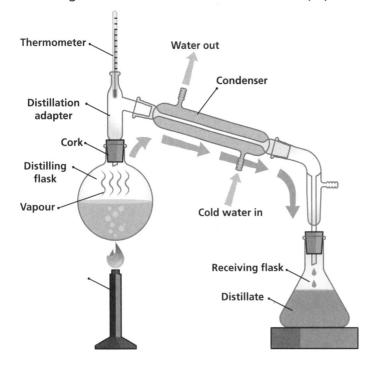

a) Describe what happens to the salt and water mixture as it is heated in the distilling flask.

_____

☐ 2 marks

b) Predict the temperature shown on the thermometer.

_____°C

1 mark

c) Describe what happens in the condenser.

_____

_____

2 marks

d) Explain why cold water flows through the outside of the condenser.

_____

1 mark

TOTAL

6

3 **a)** A group of students is investigating the way that a spring stretches when a load is added to it. They hang a spring from a hook and add weights to it.

Put a tick (✓) next to the statements that are true.

A If they double the weight on the spring, its length will double as long as the elastic limit is not exceeded.

B If they double the weight on the spring, its extension will double as long as the elastic limit is not exceeded.

C If the elastic limit of the spring is exceeded and the weights removed, the spring will not return to its original length.

D The extension of the spring is calculated by adding the original length of the spring to its length when loaded.

2 marks

b) This spring is being loaded. The picture shows the spring before being loaded, with a 10N load attached and with a 20N load attached. The length of the unloaded spring is 20cm; with 10N it is 23cm and with 20N it is 26cm.

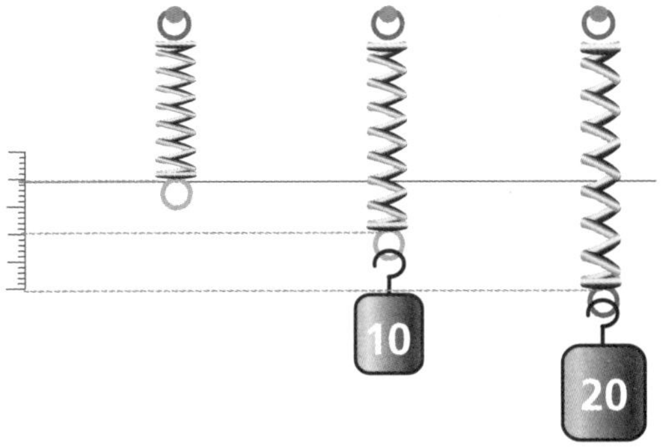

i) Calculate the extension caused by the 10N load.

_____

2 marks

Mixed Test-Style Questions

ii) Assuming the elastic limit is not exceeded, calculate the extension caused by a 40N load.

....

2 marks

iii) If a 50N load causes the length of the spring to be 35cm, has the elastic limit been exceeded?

....

1 mark

iv) If a 100N load causes the length of the spring to be 55cm, has the elastic limit been exceeded?

....

1 mark

c) A group of students are investigating the loading of a spring. They set the spring up, add loads and calculate the extension of the spring. Their data is shown in the table.

Load (N)	0	1	2	3	4	5	6	7	8
Extension of spring (mm)	0	4	9	13	18	22	29	38	50

i) Identify the dependent variable and the independent variable.

....

....

2 marks

ii) Plot a line graph to show the relationship between load and extension. Draw a line of best fit.

4 marks

iii) Use the graph to predict what extension would be caused by a load of 2.5N.

....

2 marks

iv) Explain why the line of best fit is not straight over its whole length.

....

....

2 marks

v) Identify on the graph the section which obeys Hooke's Law.

....

2 marks

vi) State the load beyond which the elastic limit was exceeded.

....

2 marks

TOTAL

22

4 **a)** Draw a plant cell and add labels to show the following parts of the cell:

cell wall **chloroplast** **cytoplasm** **cell membrane** **mitochondria** **nucleus** **vacuole**

7 marks

b) Amoeba are a type of unicellular organism.

Complete the table below. Add one tick (✓) to each row to show if the statements about unicellular organisms are true or false.

Statement	True	False
A unicellular organism is made of only one cell.		
Unicellular organisms cannot move.		
Unicellular organisms are classed as non-living things.		
You need a light microscope to view most unicellar organisms.		
Unicellular organisms have adaptations to help them survive and function in their environments.		
Substances can move in and out of unicellular organisms by diffusion.		

6 marks

c) Put the following words in their correct order to show levels of organisation, from smallest to largest. The first has been done for you.

cell   **organ**   **organism**   **organ system**   **tissue**

cell ....

3 marks

Mixed Test-Style Questions

d) Draw one line from each word to the correct definition.

Word	Definition
Pollination	The process of the embryo developing in the womb
Fertilisation	The transfer of pollen from a male part of a plant to a female part of a plant
Gestation	When male and female sex cells join

2 marks

e) Asters are a type of plant that produce flowers that contain nectar. Explain how nectar helps increase the chance of pollination in Asters.

....

....

2 marks

f) Wind-pollinated flowers do not have large, brightly coloured and scented flowers or nectar. Explain why.

....

1 mark

TOTAL

21

---

5 This question is about elements and compounds.

a) Substances can be represented using diagrams. Consider the diagrams below. The letters do not represent the symbols of the substances.

A   B   C   D

Give the letter/letters of the substances that represent:

i) an element: ....

ii) a mixture of elements: ....

iii) a compound: .... and ....

iv) a substance that could be water, H_2O: ....

4 marks

b) Compounds can be represented by their chemical name or by their chemical formula. Name the following compounds.

i) MgO

ii) $CaBr_2$

iii) CaS

iv) $CaSO_4$

4 marks

TOTAL

8

6 The periodic table shows the elements. The letters shown in the table below represent elements but are not the symbols of these elements.

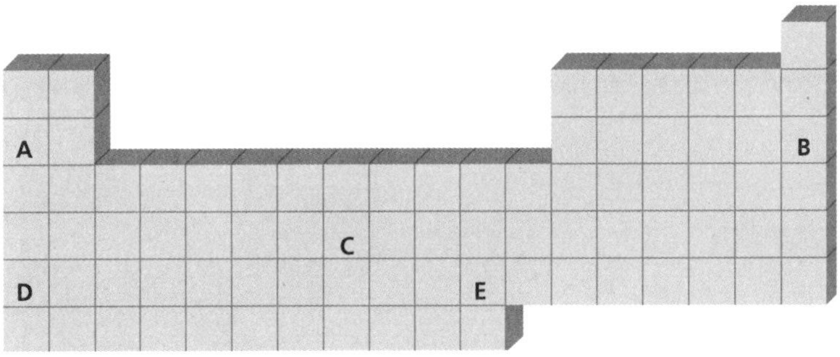

a) Explain why carbon dioxide, CO_2 is not on the periodic table.

1 mark

b) Name the vertical columns in the periodic table.

1 mark

c) Name the horizontal rows in the periodic table.

1 mark

d) Identify the letter representing the element that is a non-metal.

1 mark

e) Identify the letter representing the element that has similar chemical properties to element A.

1 mark

TOTAL

5

7 A student has a snack consisting of a glass of apple juice, a slice of toast and a banana.

Food	Mass/g	Energy Content/kJ
Apple juice	250	500
Toast	30	300
Banana	150	600

a) State which of these items has the greatest energy content.

....

1 mark

b) Calculate the energy content per gram of each of the types of food.

....

....

3 marks

c) State which type of food has the greatest energy content per gram.

....

1 mark

TOTAL

5

8 Aerobic respiration is a chemical reaction that uses glucose and oxygen.

a) Where does aerobic respiration take place? Put a tick (✓) in one box.

A Only in the muscles ☐   **B** Only in the lungs ☐

C Only in the heart ☐   **D** In all cells ☐

1 mark

Fermentation is a type of anaerobic respiration carried out by yeast.

Some students investigated the effect of temperature on fermentation in yeast using the apparatus shown below.

water   bubbles

active yeast
and glucose

The students counted the number of bubbles produced at different temperatures.

b) Name the independent variable in this investigation.

....

1 mark

c) Name one variable that the students would need to control.

....

1 mark

d) Explain why glucose was added to the yeast and warm water.

....

1 mark

e) Name the gas that forms the bubbles.

....

1 mark

f) The diagram shows the elbow joint in the arm.

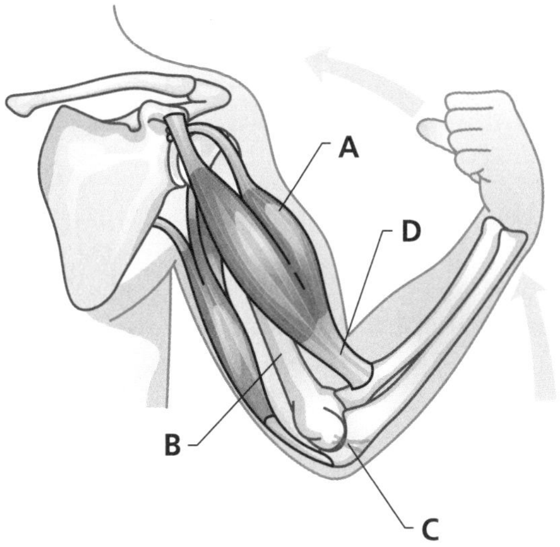

State the letter on the diagram that shows:

i) A muscle ....

1 mark

ii) A tendon ....

1 mark

iii) A ligament ....

1 mark

g) Describe how the muscle shown in the diagram bends the arm.

....

....

2 marks

h) Unlike animals that eat food, plants make their own glucose during photosynthesis. Describe how plants make their own glucose.

....

....

....

4 marks

TOTAL

14

9 Which of these processes removes carbon dioxide from the atmosphere? Put a tick (✓) in one box.

A Respiration by plants ☐

B Respiration by animals ☐

C Photosynthesis by plants ☐

D Burning fossil fuels ☐

1 mark

TOTAL

1

10 This question is about pressure.

a) The diagram shows a container of water with several holes in it. The holes are at different heights and water is running out of the holes into a tray.

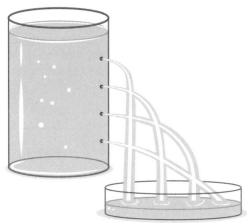

i) Explain what this shows about the water pressure in the container at different depths.

....

....

2 marks

ii) Explain why the pressure varies in this way.

2 marks

iii) Suggest what would be observed if another hole was made in the container, the same height as the lowest hole but on the other side.

1 mark

iv) If this experiment was left running for a few minutes, the water level in the container would gradually drop. Suggest what would happen to the streams of water from the holes and explain why.

2 marks

b) A submarine is diving under the water in the sea.

i) Explain why the pressure on the submarine gets greater the deeper it dives.

2 marks

ii) Explain what would happen if it dived too deep.

1 mark

c) The air pressure at the top of a tall mountain is lower than that at sea level.

i) Explain why this is the case.

2 marks

ii) Some mountaineers are exploring differences in pressure. When they were at the foot of the mountain, they partially inflated a balloon with air and tied a knot in it. They took the balloon up the mountain with them. The higher they climbed, the larger the balloon got. Explain why this happened.

3 marks

TOTAL

15

11 The picture shows the expansion gap in a road. The surface to the left of the gap is part of a bridge; the surface to the right is where the road is laid on the ground.

a) Describe what will happen to the length of the bridge on a hot day.

1 mark

b) Describe what will happen to the size of the expansion gap on a hot day.

1 mark

c) Describe what would happen to the road and the bridge if there was no expansion gap.

2 marks

TOTAL

4

12 This question is about chemical reactions.

A teacher dissolves a sample of ammonium nitrate, NH_4NO_3 into a beaker of water. At the start of the experiment the temperature of the water was 20.0°C. When the ammonium nitrate was added to the beaker, the temperature of the contents of the beaker changed to 18.9°C.

a) Identify the three elements in ammonium nitrate.

_____

☐ 1 mark

b) Complete the table by putting a tick (✓) in the correct column to show whether each of the substances involved in the experiment is pure or impure.

Substance	Pure	Impure
Ammonium nitrate		
Water		
Solution of ammonium nitrate		

☐ 3 marks

c) Calculate the temperature change in this experiment.

_____

☐ 1 mark

d) Identify the type of reaction that takes place in this experiment. Put a tick (✓) in one box.

A Exothermic ☐   B Endothermic ☐

C Neutralisation ☐   D Thermal decomposition ☐

☐ 1 mark

e) Complete the diagram to show how the energy moves from one store to another store during this experiment.

_____ store ⟶ _____ store

☐ 2 marks

TOTAL

☐ **8**

13 A student takes an ice cube out of the freezer. She finds the ice cube has a mass of 2.8g.

a) Name the piece of equipment the student uses to find the mass of the ice cube.

_____

☐ 1 mark

b) The student leaves the ice cube in a warm room until it changes into liquid water. Name the change in state that has happened.

_____

☐ 1 mark

Mixed Test-Style Questions

c) Predict the mass of the liquid water at the end of the experiment.

_____ g

1 mark

TOTAL

3

14 A student digs up an old nail in his garden. He notices that the iron is covered in rust.

a) State the conditions required for iron to rust.

_____

2 marks

b) The metal zinc also corrodes. Explain why zinc corrodes faster than iron.

_____

1 mark

c) Name a metal that corrodes more slowly than iron.

_____

1 mark

TOTAL

4

15 A student adds some granules of zinc metal to a beaker of dilute hydrochloric acid. He observes bubbles of hydrogen gas being made. A solution of zinc chloride is also produced.

a) Write a word equation for this reaction.

_____

1 mark

b) The student then adds a gold ring to the acid. Describe and explain what would happen to the ring.

_____

_____

2 marks

TOTAL

3

16 This diagram shows how light rays are focused by the eye to form an image.

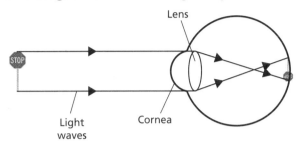

a) Explain the function of the lens in the eye.

....

....

1 mark

b) Describe the size of the image compared with the size of the object.

....

1 mark

c) Explain why the image formed is inverted.

....

1 mark

d) This diagram shows how a pinhole camera forms an image.

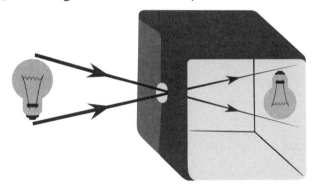

State two similarities and two differences between the way that a pinhole camera forms an image and the way the human eye forms an image.

....

....

....

....

4 marks

TOTAL

7

Mixed Test-Style Questions

17 **a)** These questions are about the reflection of white light.

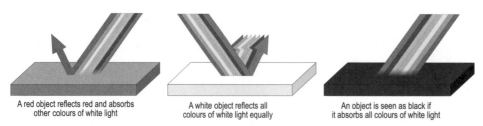

A red object reflects red and absorbs other colours of white light

A white object reflects all colours of white light equally

An object is seen as black if it absorbs all colours of white light

i) Describe an experiment you could carry out to show that white light consists of a number of different colours of light.

4 marks

ii) Describe what would be reflected and what would be absorbed if white light was to land on a blue object.

2 marks

iii) Suggest why scientists may refer to white light, red light, etc., but not black light.

2 marks

b) This question is about coloured filters.

i) What colour light will pass through a blue filter? Put a tick (✓) in the box next to the correct answer.

A Red ☐   **B** Blue ☐   **C** Green ☐   **D** White ☐

1 mark

ii) Which colour in white light is **not** absorbed by a red filter? Put a tick (✓) in the box next to the correct answer.

A Red ☐   **B** Blue ☐   **C** Green ☐   **D** White ☐

1 mark

TOTAL

☐

10

18 **a)** Plants are adapted to carry out photosynthesis.

Draw a line from each part of the plant to its correct function.

Part of the Plant	Function
Stomata	Transport water and minerals from the roots to the leaves
Xylem	To allow gases to move in and out of the leaf
Phloem	Transport glucose around the plant

2 marks

b) Giant rhubarb plants have leaves that grow up to 130cm in diameter.

Explain why it is an advantage for the giant rhubarb plant to have such large leaves.

....

2 marks

The diagram below shows four living things found in a grassland habitat.

snake   grass   hawk   grasshopper

- Grasshoppers eat grass.
- Snakes eat grasshoppers.
- Hawks eat snakes.

c) **i)** Draw a food chain for these organisms.

2 marks

ii) Name the organism that is the producer.

....

1 mark

iii) Name an organism that is a consumer.

....

1 mark

Species of hawks have common features. Hawks reproduce by sexual reproduction.

The colours of the feathers are similar, but there is slight variation between individual hawks, even within a species.

d) i) State the scientific term used to describe variation within a species.

☐ 1 mark

ii) Explain the causes of variation in feather colouring in hawks.

☐ 2 marks

TOTAL

☐ 11

19 Propane, C_3H_8 is a fuel that is used in camping stoves.

a) Name the two elements found in a molecule of propane.

_____ and _____

☐ 2 marks

b) Name the gas that propane reacts with when it is burnt in the camping stove.

☐ 1 mark

c) When propane is burnt, carbon dioxide gas is produced. Name one other product of the complete combustion of propane.

☐ 1 mark

d) Explain why scientists are concerned about the rising levels of carbon dioxide in the Earth's atmosphere.

☐ 1 mark

TOTAL

☐ 5

20 Three students are investigating what happens when salt dissolves in water. They measure the mass of water in a beaker and then dissolve a measured mass of salt in the water. They each make a prediction about what the mass of the salt solution will be:

Archie: I think the mass of the salt solution will be less than the total mass of the water and the salt because when salt dissolves it disappears.

Belinda: I think that the mass of the solution will be equal to the mass of the water plus the mass of the salt because mass is conserved when a solid dissolves in a liquid.

Suhaib: I think the mass of the solution will be greater than the mass of the water plus the mass of the salt because salt water is denser than pure water.

Comment on each prediction saying whether you think it is correct or incorrect and why.

6 marks

TOTAL

6

21 Wasp stings are very painful. The venom in the wasp's sting is alkaline. Explain why applying vinegar onto the site of a wasp sting might make the sting less painful.

2 marks

TOTAL

2

Mixed Test-Style Questions

22 This piece of equipment is a pile driver. It is used to force pillars into the ground to lay the foundations for buildings. The large weight at the top left of the picture is raised up by the engine and then allowed to drop. When it lands, it hits the pillar and pushes it deeper into the ground. It makes quite a lot of noise.

a) As the weight is being raised, is its gravitational potential store increasing, decreasing or staying the same?

....

1 mark

b) As the weight falls, is its gravitational potential store increasing, decreasing or staying the same?

....

1 mark

c) As the weight falls, is its kinetic store increasing, decreasing or staying the same?

....

1 mark

d) When the weight has struck the pillar and pushed it down a bit, where has the energy that was in the weight then been transferred to?

....

1 mark

TOTAL

4

23 The picture shows a simple electromagnet. A student has set this up and tested it. He knows from using a compass that the right-hand end is the North pole.

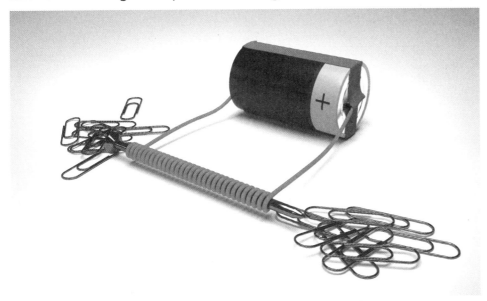

a) Draw a simple diagram of the electromagnet and add details to show the magnetic field around the coil.

2 mark

b) Describe what the student could do to reverse the poles of the electromagnet.

1 mark

TOTAL

3

Mixed Test-Style Questions

24 Decide which of these statements are true and which are false. Put a tick (✓) in the correct column.

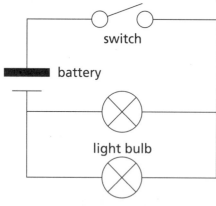

Parallel Circuit

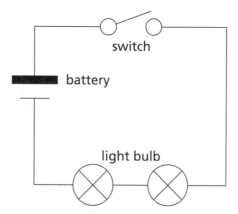

Series Circuit

	True	False
a) In this parallel circuit both of the bulbs have the same potential difference across them as that supplied by the battery.		
b) In a series circuit, the same current flows at all points in the circuit.		
c) In this parallel circuit, if one bulb blew the other would go out too.		
d) In this series circuit, adding more bulbs in the loop would not affect the brightness of the other bulbs.		

4 marks

TOTAL

4

Answers

Pages 148–149 Cells – the Building Blocks of Life

1. a) A – Nucleus [1]; B – Cell membrane [1]
   b) It controls what substances can enter and leave the cell. [1]

2.

Name of Structure	Function
Cell wall	**Gives rigid support to the cell**
Cytoplasm	Place where lots of chemical reactions (photosynthesis) take place
Chloroplasts	**Absorbs light energy to allow plants to make their own food.**
Vacuole	Contains cell sap and keeps the cell firm.

[4]

3.

Description of Step	Order (1–5) Starting with 1
Place the slide on the stage of the microscope.	4
Stain the sample.	2
Illuminate and focus the microscope.	5
Place the cell sample on a slide.	1
Cover the sample with a coverslip.	3

[5]

4. a) Because there is a higher concentration of glucose molecules outside of the cell and so they will diffuse into the area of lower concentration of glucose inside the cell. [1]

b) **Accept two from:** Because it does not have chloroplasts; cell wall; large vacuole [2]

5. a) Pollen is transferred from the male part of the plant/anther [1] to the female part of the flower/stigma. [1]
   b) The nectar attracts insects to fly from one flower to another. [1] This increases the chances of pollen transfer/pollination. [1]

6. Male – Testes [1]
   Female – Ovaries [1]

7. Correct order: fertilisation, gestation, birth [3]

Pages 150–151 Eating, Drinking and Breathing

1. a) A – Trachea [1]; B – Bronchiole [1]; C – Alveoli [1]; D – Lung [1]
   b) Gas exchange [1]

2. When we inhale, the ribcage moves upwards and **outwards**. [1] The diaphragm **contracts** [1] and flattens. This **increases** [1] the volume of the chest cavity and **decreases** [1] the pressure, causing air to move into the lungs from outside.

3. The airways/bronchioles constrict/get narrower [1] making breathing harder/more difficult. [1]

4. Damaged cilia causes mucus to start to build up in the airways, making it harder for the smoker to breathe [1] so the smoker has to cough to get rid of the mucus. [1]

5. Carbohydrates – Provide energy [1]; Proteins – Important for growth and repair [1]; Fats – Provide a reserve energy supply and insulation [1]; Fibre – Helps undigested food pass quickly through the gut [1]

6. oesophagus before stomach [1]; stomach before small intestine [1]; small intestine before large intestine [1]; large intestine before rectum [1]; rectum before anus [1]

7. Because anaemia is caused by a lack of iron [1] and these foods are rich in iron. [1]

8. a) They are used to break up larger food molecules into smaller ones [1] so they can pass through the intestine wall. [1]
   b) Protein ✓ [1]

Pages 152–153 Getting the Energy your Body Needs

1.

Statement	True	False
Respiration is a chemical reaction.	✓	
Respiration takes place in the cells of plants and animals.	✓	
Respiration only takes place in the lungs.		✓
Respiration releases energy.	✓	

[4]

2. Oxygen [1]; Carbon dioxide [1]

3. a) Glucose [1] → Lactic acid [1]
   (It does also produce energy but as this is not a substance, it is not a product in the word equation.)
   b) **Any three from:** Aerobic uses oxygen as a reactant whereas anaerobic does not; Aerobic produces carbon dioxide and water, but anaerobic does not; Anaerobic produces lactic acid but aerobic does not; Aerobic releases more energy than anaerobic; Both require glucose; Both release energy. [3]

4. To provide support ✓ [1]
   To protect organs ✓[1]
   To make red blood cells ✓[1]
   To allow the body to move ✓ [1]

5. a)

Muscle	Bending the Leg	Straightening the Leg
Quadriceps	Relaxes	Contracts
Hamstring	Contracts	Relaxes

[4]

   b) Because it provides a smooth surface that cushions the joint/reduces friction in the joint/protects the joint. [1]
   c) A tendon attaches a muscle to bone. [1] A ligament attaches bone to bone. [1]

Pages 154–155 Looking at Plants and Ecosystems

1. a) light [1]; glucose [1]
   b) Carbon dioxide [1]; Glucose [1]

2.

Adaptation of Leaf	Function
Are flat and broad	To give a large surface area for absorbing light.
Lower surface contains stomata	**To control the diffusion of gases in and out of the leaf.**
Contain xylem and phloem tubes	**To transport water and glucose.**

[2]

3. a) They allow gases/carbon dioxide and oxygen to move in and out of the leaf. [1]
   b) The stomata allow carbon dioxide to enter the leaf so that photosynthesis can take place [1] and the plant can make its own food. [1]

4. a) **Any two from:**
   Algae → Tadpole → Small fish → Kingfisher
   Algae → Water beetle → frog → Kingfisher
   Algae → Snail → Frog → Kingfisher [2]

b) The algae absorbs the mercury from the pond. **[1]**

The mercury is then passed on to the animals that consume/eat the algae. **[1]**

Because the animals eat a lot of the algae, they absorb more of the poison. **[1]**

The poison accumulates as it is passed up the food web, killing the kingfisher. **[1]**

5. Interdependency ✓ **[1]**

Pages 156–157 Variation for Survival

1.

Statement	True	False
A baby inherits half its genetic information from each parent.	✓	
Brothers and sisters inherit the same genetic information from their parents.		✓ (except identical twins)
Genetic information is stored within chromosomes, found within the nucleus of cells.	✓	
DNA is only found in egg and sperm cells.		✓

[4]

2. **A:** Chromosome **[1]**; **B:** DNA/DNA helix **[1]**; **C:** Gene **[1]**

3. Because the egg and sperm cell join in fertilisation **[1]** to make a new cell/baby with 46 chromosomes. **[1]**

4. Foot length – Continuous **[1]**; Blood group – Discontinuous **[1]**; Hamster weight – Continuous **[1]**; Fur colour – Discontinuous **[1]**

5. Because more variation within a species means that there is more chance some individuals in the species will be better adapted to an environmental change **[1]** and more likely to survive and reproduce. **[1]**

6. **Any two from:** To prevent species becoming extinct; To potentially provide new foods/medicines in the future; To maintain species biodiversity **[2]**

7. Watson and Crick – Developed a theory for the structure of DNA **[1]**; Maurice Wilkins – Produced evidence to support Watson and Crick's theory **[1]**; Rosalind Franklin – Made X-ray images of DNA **[1]**

Pages 158–159 Our Health and the Effects of Drugs

1.

Statement	True	False
A drug is a chemical that alters the way your body or mind works.	✓	
Only a few drugs cause side effects.		✓
Caffeine, alcohol and nicotine are all drugs.	✓	
Only illegal drugs are additive.		✓

[4]

2. Stimulant – Makes a person feel energetic and alert. **[1]**; Depressant – Makes a person feel relaxed and drowsy. **[1]**; Painkiller – Reduces pain and inflammation. **[1]**; Hallucinogen – Makes a person hear or see things that are not real. **[1]**

3. **Two from:** Liver damage/cirrhosis of liver; Brain damage; Increased risk of strokes and heart attacks; Anxiety and depression; Mental health issues **[2]**

4. Bacteria **[1]**; Viruses **[1]**; Fungi **[1]**

5.

Part of Body	Defence
Eyes	Produce tears that contain a chemical to kill microbes
Ears	**Produce wax to trap microbes.**
Nose and throat	**Produce mucus to trap microbes.**

[2]

6. **Any two from**: They surround, engulf and destroy microbes; They produce antibodies to destroy the pathogen; They produce chemicals/antitoxins to neutralise toxins made by microbes; [2]

7. a) Dead microbes are injected into the body [1] causing the blood to make memory cells that offer protection against that microbe. [1]

   b) Because Covid-19 is caused by a virus [1] and antibiotics are only effective against bacteria, not viruses. [1]

Pages 160–171 Chemistry

Pages 160–161 Mixing, Dissolving and Separating

1. Orange juice ✓ [1]
2. a) C [1]
   b) A and D [1]
   c) It is insoluble (in this solvent). [1]
   d) E. [1] It has the most spots/It has three spots. [1]
3. **A** Residue [1]
   B Filtrate [1]
   C Filter funnel [1]
   D Filter paper [1]
4. a) Magnesium oxide [1]
   b) 2.0g [1]
5. Carbon dioxide and water ✓ [1]

Pages 162–163 Elements, Compounds and Reactions

1. Atom – A very small particle; Element – Contains only one type of atom; Compound – Contains different types of atoms joined together by chemical bonds; Group – A column in the periodic table [4]
2. a) 6 [1]
   b) 3 [1]
   c) 24 [1]

3. Copper – Cu; Silver – Ag; Magnesium – Mg; Sulfur – S; Sodium – Na [5]
4. a) Magnesium [1] and hydrochloric acid [1]
   b) Magnesium chloride [1]
   c) A gas is made. [1]
   d) Conducts electricity well ✓ [1] Shiny when cut ✓ [1] High boiling point ✓ [1]
5. a) Copper oxide ✓ [1]
   b) Oxidation ✓ [1]

Pages 164–165 Explaining Physical Changes

1. a) i) Solid [1] and liquid [1]
      ii) Gas [1]
      iii) Solid [1]
      iv) Liquid [1]
      v) Gas [1]
   b) Sublimation [1]
2. The dust particles are hit by (smaller) air particles. [1] The air particles move about randomly/move the dust particles. [1]
   Accept Brownian motion **for [1]**
3. The rubbish evaporates/turns into a gas. [1] These molecules diffuse through the air. [1]
4. a) (Type of) metal [1]
   b) Temperature change [1]
   c) There is no reaction [1]
   d) + 5.0 [1]
5. Sublimation ✓ [1]

Pages 166–167 Explaining Chemical Changes

1. a)

Everyday Substances	pH Number	Colour with U.I.	Type of Solution
Lemon juice	2	**Red/ orange**	**Acid**
Hair dye	11	**Blue/ purple**	**Alkali**
Seawater	7	Green	**Neutral**
Lemonade	**(3 to 6)**	Orange	**Acid**

[4]

   b) i) protect her eyes [1]
      ii) protect her skin/hands [1]
2. a) Sodium sulfate [1]
   b) Calcium chloride [1] carbon dioxide [1]

3. a) Red [1]

   b) It tells you how acidic (or alkaline) a solution is. [1]

4. Increases the rate of reaction. ✓ [1]

5. a) (g) [1] (g) [1] (l) [1]

   b) Aqueous [2] or dissolved [1] in water [1]

6. $C_2H_6 + 3\frac{1}{2}O_2 \rightarrow 2CO_2 + 3H_2O$ ✓ [1]

7. a) 1 ✓ [1]

   b) Yes [1] Carbon dioxide/a gas is made. [1]

Pages 168–169 Obtaining Useful Materials

1. Composite – A material that is made from two or more different materials joined together; Polymer – A material made by chemically joining together many smaller molecules; Monomers – Small molecules; Displacement – When a more reactive metal takes the place of a less reactive metal in its compound [4]

2. a) (Most reactive) magnesium, zinc, iron, copper (Least reactive) [1]

   b) magnesium + hydrochloric acid [1] → magnesium chloride + hydrogen [1]

   c) Magnesium is more reactive than carbon. [1]

3. magnesium + copper sulfate → magnesium sulfate + copper [1]

4. a) (Most reactive) C, D, B, A (Least reactive) [1]

   b) A [1]

Pages 170–171 Using our Earth Sustainably

1. a) Sedimentary – A soft rock which may contain fossils; Metamorphic – A hard rock which often has layers; Igneous – A very hard rock with interlocking crystals [3]

   b) To make igneous rock, the rock is melted [1] so the remains would be destroyed/melt. [1]

   c) High pressure and high temperatures ✓ [1]

2. a) Nitrogen [1]

   b) Oxygen [1]

3. a) 135 [1]

   b) Burning petrol in cars ✓ [1]

4. **Any two of:** Less mining for raw materials; Less landfill; Less rubbish; Fewer fossil fuels burnt; Less carbon dioxide made; Less damage to wildlife. [2]

5. Crust ✓ [1]

6. a) Plants [1] carry out photosynthesis. [1]

   b) Plants respire [1], returning carbon as carbon dioxide. [1]

Pages 172–183 Physics

Pages 172–173 Forces and their Effects

1. a) They are equal in size [1] and opposite in direction. [1]

   b) They are equal in size. [1]

   c) T must be greater than F. [1]

   d) Streamlining reduces friction (or drag) which means less thrust is needed to accelerate the car and keep it moving so less fuel is needed. [1]

2. a) The small rock under the pole. [1]

   b) The force on the left-hand end of the pole is less than the force applied to the boulder. [1]

   c) Move the fulcrum [1] towards the boulder so that it is closer. [1]

3. a) Turning effect (or moment) = force × distance [1]

   b) Newtonmetres [1]

   c) Turning effect = 50N × 0.75m = 37.5Nm [1]

4. a) Moment = 50N × 3.2m = 160Nm [1]

   b) They must be equal in size [1] and opposite in direction. [1]

   c) Distance = 160Nm ÷ 200N = 0.8m [1]

Pages 174–175 Exploring Contact and Non-Contact Forces

1. a) On the slide, where the child is sliding down, [1] so they go faster. [1]

   b) On the steps, [1] so their feet don't slip. [1]

2.

Force Acting	Contact Force	Non-contact Force
A person pushing down on a door handle.	✓	
A charged rod attracting bits of tissue paper.		✓
An apple falling towards the ground.		✓
A match being struck on a matchbox.	✓	

[4]

3. Gravity is a **force [1]** which attracts masses towards each other. The space in which **gravity [1]** acts on objects is called the gravitational **field. [1]** The strength of a gravitational **field [1]** is measured in **Newtons [1]** per kilogram (N/kg).

4. a) 40 × 40 **[1]** = 1600cm² **[1]**
   b) 200 ÷ 1600 **[1]** = 0.125N/cm² **[1]**
   c) 5 × 40 **[1]** = 200cm² **[1]**
   d) 200 ÷ 200 **[1]** = 1N/cm² **[1]**
   e) The surface area is much smaller **[1]** so the force is more concentrated. **[1]**

5. a) Ice is less dense than water. **[1]**
   b) With most materials the density of the solid form is greater than that of the liquid form. **[1]**

Pages 176–177 Motion on Earth and in Space

1.

Object	Distance Travelled	Time Taken	Speed
Cyclist	100m	**10s**	10m/s
Train	**4800m**	2 minutes	40m/s
Sound wave	**3300m**	10s	330m/s
Runner	400m	**100s**	4m/s

[4]

2. a) moving at a steady speed ✓ **[1]**
   b) 40m ÷ 4s = 10m/s **[1]**
   c) Steady speed, followed by lower steady speed, followed by steady speed greater than the second section but less than the first section. **[1]**
   d) The first section is 20m ÷ 1s = 20m/s, **[1]** the second section is 10m ÷ 2s = 5m/s, **[1]** and the third section is 10m ÷ 1s = 10m/s **[1]**
   e) 40m ÷ 4s = 10m/s **[1]**

3. a) When a hemisphere is tilted towards the Sun, the days will be longer and the Sun higher in the sky at midday; it will be summer. **[1]** When tilted away from the Sun, the days will be shorter and the Sun lower in the sky; it will be winter. **[1]**
   b) i) Summer **[1]**
      ii) Winter **[1]**
   c) There would be less seasonal variation. The difference in day length and temperatures between summer and winter would be less, **[1]** so summers wouldn't be as hot and winters wouldn't be as cold. **[1]**

4. a) 100kg × 10N/kg **[1]** = 1000N **[1]**
   b) 100kg × 1.7N/kg **[1]** = 170N **[1]**
   c) 100kg × 8.9N/kg **[1]** = 890N **[1]**

Pages 178–179 Energy Transfers and Sound

1.

Fuel	Renewable	Non-renewable
Coal		✓
Wind	✓	
Nuclear		✓
Natural gas		✓
Solar	✓	
Oil		✓

[6]

2. a) 2000W **[1]**
   b) 2000J/s **[1]**
   c) 2000 × 60 = **[1]** 120,000J **[1]**
   d) Some of the energy is transferred into the kettle itself and some into the surroundings. **[1]**

3. a) Three times greater [1]
    b) **Answers could include:** They might be able to find it locally; It might be cheaper or free; It is safer; The fire is more attractive. [2]
    c) **Answers could include:** It is more convenient; It is lighter to carry; It is a more concentrated source of energy; It is easier to light and control. [2]

4. Sounds are made when something vibrates, such as the **skin [1]** of a drum or the **strings [1]** of a guitar. These **vibrations [1]** travel through the air because of particles vibrating. Sound needs a medium to **travel [1]** through. We can hear **sound [1]** because the vibrations transfer **energy [1]** to our ears.

5. a) 20,000Hz [1]
    b) The range of frequencies that can be heard. [1]
    c) They have a frequency greater than 20kHz. [1]

6. a) The speed of sound is much slower than the speed of light. [1]
    b) There would be less of a pause between seeing the point of contact and hearing it. [1]

7. Energy has been transferred from the ice cubes to the water. ✓
    The internal energy of the water is less than it was before the ice was added. ✓ [2]

Pages 180–181 Magnetism and Electricity

1. a) True [1]
    b) False [1]
    c) True [1]
    d) True [1]

2. a) 0.2A; it will be the same as A_1. [1] The same current returns to the battery as leaves it.[1]
    b) 0.1A [1]; The current divides between the two loops. [1] The 0.2A is shared and as 0.1A goes through the upper loop there must be 0.1A in the lower loop. [1]

3. a) 2.8V [1]; The voltage is shared between the bulbs so is calculated from 6.0 – 3.2. [1]
    b) It shows that the bulb on the left [1] has a greater resistance. [1]

4. When opposite poles are brought near to each other they attract [1] but when like poles are brought close they repel. [1]

5. 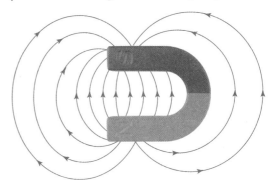 [2]

6. She can detect the magnets because they will be attracted by one end of her magnet and repelled by the other end. [1] She can detect the magnetic materials because they will be attracted by either end of her magnet [1] and she can detect the non-magnetic materials because they won't be attracted or repelled by her magnet. [1]

7. Replace the wooden rod with an iron core; Add more coils; Increase the current flowing (by increasing the voltage). [3]

Pages 182–183 Waves and Energy Transfer

1. a) B [1]
    b) C [1]
    c)

    Peak [1]

    Trough [1]

2. a) Hertz (Hz) [1]
    b) Metres (m) (**Also accept:** centimetres (cm); millimetres (mm); kilometres (km)) [1]
    c) Metres (m) (**Also accept:** centimetres (cm); millimetres (mm)) [1]
    d) Seconds [1]

3. a) Longitudinal = B **[1]**; Transverse = A **[1]**
    b) Longitudinal – energy is transferred to the wave in the same direction as the direction in which the wave travels. **[1]** Transverse – energy is transferred to the wave at right angles to the direction of wave travel. **[1]**

4. a) The normal **[1]**
    b) They are equal in size. **[1]**
    c) The light appears to travel from there **[1]** but is actually travelling from the object and being reflected by the mirror. **[1]** The line is dotted to show that light isn't actually travelling along that route (it is virtual rather than real). **[1]**

5. a) Refraction ✓ **[1]**
    b) It makes the rays close in and meet. **[1]**
    c) The rays would still close in and meet at a focal point. **[1]** However, they wouldn't be refracted by as much so the focal point would be further from the lens. **[1]**

> **Pages 184–200 Mixed Test-Style Questions**

1. a) **Any one of:** touching; close together; uniformly; regularly **[1]**
    b) A gas is made. **[1]**
    c) Filter off the unreacted calcium carbonate. **[1]** Collect the filtrate. **[1]** Warm the mixture. **[1]**

2. a) The water **[1]** evaporates/boils. **[1]**
    b) 100°C **[1]**
    c) The water vapour/gas **[1]** turns back into a liquid/condenses. **[1]**
    d) To cool the gas. **[1]**

3. a) A ✓ B ✓ **[2]**
    b) i) 3cm **[2]**
       ii) 12cm **[2]**
       iii) No **[1]**
       iv) Yes **[1]**
    c) i) The load is the independent variable **[1]** and the extension is the dependent variable. **[1]**

   ii) **Award [1] for axes drawn correctly; [1] for axes calibrated; [1] for all points plotted correctly; [1] for line of best fit drawn accurately.**

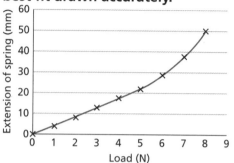

   iii) 11mm **[2]**
   iv) Because the spring passes its elastic limit **[1]** and starts to become permanently deformed. **[1]**
   v) The section from 0N **[1]** to 5N **[1]**
   vi) 5N **[2]**

4. a) **Award [1] per correct label up to [7].**

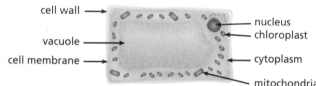

b)

Statement	True	False
A unicellular organism is made of only one cell.	✓	
Unicellular organisms cannot move.		✓
Unicellular organisms are classed as non-living things.		✓
You need a light microscope to view most unicellar organisms.	✓	
Unicellular organisms have adaptations to help them survive and function in their environments.	✓	
Substances can move in and out of unicellular organisms by diffusion.	✓	

[6]

c) tissue before organ [1]; organ before organ system [1]; organ system before organism [1]

d) [1] for one correct answer and [2] for three correct answers.
Pollination – The transfer of pollen from a male part of a plant to a female part of a plant
Fertilisation – When male and female sex cells join
Gestation – The process of the embryo developing in the womb

e) The nectar attracts insects to visit the Aster flowers. [1] When these insects fly from one Aster to another, they carry the pollen which increases the chance of pollination. [1]

f) Because they do not need to attract insects. [1]

5. a) i) B [1]
   ii) A [1]
   iii) C and D [1]
   vi) D [1]

b) i) magnesium oxide [1]
   ii) calcium bromide [1]
   iv) calcium sulfide [1]
   v) calcium sulfate [1]

6. a) It is a compound/is not an element. [1]
   b) Groups [1]
   c) Periods [1]
   d) B [1]
   e) D [1]

7. a) Banana [1]
   b) Apple juice: 2kJ/g [1] Toast: 10kJ/g [1] Banana: 4kJ/g [1]
   c) Toast [1]

8. a) D ✓ [1]
   b) Temperature [1]
   c) **Any one from:** The same batch of yeast; The same amount of yeast; The amount of glucose; The same apparatus. [1]
   d) Because glucose is a reactant in fermentation. [1]

e) Carbon dioxide [1]
f) i) A [1]   ii) D [1]   iii) C [1]
g) The muscle contracts/shortens, [1] which pulls the lower arm bones and moves the joint. [1]
h) They take carbon dioxide [1] and water [1] and use energy from the Sun [1] to convert these substances into glucose. [1]

9. C ✓ [1]

10. a) i) The pressure is more [1] at a great depth. [1]
    ii) At the greater depth, there is more water above that point [1], pressing down. [1]
    iii) The water would spurt out in that direction but landing the same distance away from the container. [1]
    iv) After a few minutes the jets of water would be landing closer to the base of the container before stopping completely. [1] This is because as the water level dropped there would be less pressure at each point due to less water above the hole. [1]

b) i) There is more water above the submarine [1] so the pressure is greater. [1]
   ii) The water pressure would crush the submarine. [1]

c) i) Higher up the mountain, there is less air above that point [1] and so less air pressure. [1]
   ii) The higher they went, the pressure on the outside of the balloon became less. [1] The pressure on the inside remained the same. [1] The increasing difference in pressure caused the balloon to expand. [1]

11. a) It will increase. [1]
    b) It will decrease. [1]
    c) On a hot day the forces would build up [1] and cause the road and/or the bridge to buckle and distort. [1]

12. a) nitrogen, hydrogen and oxygen [1]

b)

Substance	Pure	Impure
Ammonium nitrate	✓	
Water	✓	
Solution of ammonium nitrate		✓

[3]

c) -1.1°C [1]

d) B ✓ [1]

e) Thermal/heat [1] Chemical [1]

13. a) Mass balance/scales [1]

b) Melting [1]

c) 2.8g [1]

14. a) Oxygen [1] and water [1]

b) It is more reactive. [1]

c) **Any one of**: gold; silver; copper; tin; lead; platinum [1]

15. a) zinc + hydrochloric acid → zinc chloride + hydrogen [1]

b) No reaction/nothing. [1] Gold is unreactive. [1]

16. a) To bend the light rays, focusing them to form an image on the retina. [1]

b) The image is smaller than the object. [1]

c) Because the light rays cross over in the eye. [1]

d) **Similarities: Any two from:** The light rays cross over to form an image; The image is inverted; Both devices only let light in through a small aperture at the front. [2]

Differences: Any two from: The eye has a lens which will bend light rays to form a focused image; The eye has a retina which converts the image into a nervous impulse; The eye can control the amount of light entering. [2]

17. a) i) Set up a triangular prism with a beam of white light in a darkened room. [1] Set up a white screen on the other side of the prism [1] and adjust until light travelling through the prism

lands on the screen. [1] The screen will then show a range of different colours of light. [1]

ii) Blue light would be reflected [1] and other colours such as red and green would be absorbed. [1]

iii) Colours of light other than black are produced by particular frequencies of light (such as red) or combinations of colours (such as white). [1] Black however, is the absence of light. [1]

b) i) B ✓ [1]

ii) A ✓ [1]

18. a) [1] for one correct line and [2] for three correct lines.

Stomata – To allow gases to move in and out of the leaf

Xylem – Transport water and minerals from the roots to the leaves

Phloem – Transport glucose around the plant

b) The large leaves provide a large surface area, [1] which means the plant can collect more light energy for photosynthesis. [1]

c) i) Grass → Grasshopper → Snake → Hawk [1] for correct order of organism and [1] for correct use of arrows.

ii) Grass [1]

iii) **Any one from:** Grasshopper; Snake; Hawk [1]

d) i) Intraspecific variation [1]

ii) Sexual reproduction between male and female hawks mixes up the genes [1] so each offspring has a unique set of genes and this causes variation in feather colours. [1]

19. a) Carbon [1] and hydrogen [1]

b) Oxygen [1]

c) Water (vapour) [1]

d) It is a greenhouse gas/climate change/ global warming. [1]

20. Archie is incorrect. **[1]** Even though the salt seems to disappear, it is still there (and can be retrieved) and its mass will add to the mass of the water. **[1]**

Belinda is correct. **[1]** Mass is conserved when a solution is formed. **[1]**

Suhaib is incorrect. **[1]** Saltwater is denser than pure water; its weight is equal to the weight of the water plus the weight of the salt (but not more than those). **[1]**

21. Vinegar is an acid. **[1]** It neutralises (the wasp venom). **[1]**

22. **a)** Increasing **[1]**
    b) Decreasing **[1]**
    c) Increasing **[1]**
    d) The surroundings, by means of thermal energy and sound. **[1]**

23. **a)**

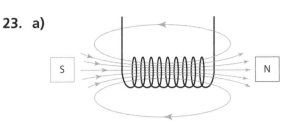

[2]

b) Reverse the connections to the battery. **[1]**

24. **a)** True **[1]**
    b) True **[1]**
    c) False **[1]**
    d) False **[1]**

ACKNOWLEDGEMENTS

The authors and publisher are grateful to the copyright holders for permission to use quoted materials and images.

Illustrations by Jouve or Shutterstock.com

Every effort has been made to trace copyright holders and obtain their permission for the use of copyright material. The authors and publisher will gladly receive information enabling them to rectify any error or omission in subsequent editions. All facts are correct at time of going to press.

Published by Collins
An imprint of HarperCollins*Publishers*
1 London Bridge Street
London SE1 9GF

HarperCollins*Publishers*
Macken House
39/40 Mayor Street Upper
Dublin 1
D01 C9W8
Ireland

© HarperCollins*Publishers* Limited 2022

ISBN 9780008551476

This edition published 2022

10 9 8 7 6 5 4

British Library Cataloguing in Publication Data.

A CIP record of this book is available from the British Library.

Authors: Byron Dawson, Eliot Attridge, Heidi Foxford, Emma Poole and Ed Walsh
Commissioning Editors: Daniel Dyer and Clare Souza
Project Managed by Beth Hutchins and Katie Galloway.
Cover Design: Kevin Robbins
Inside Concept Design: Sarah Duxbury and Paul Oates
Text Design and Layout: Jouve India Private Limited
Production: Emma Wood
Printed by Martins the Printers

FSC
www.fsc.org

MIX
Paper | Supporting
responsible forestry
FSC™ C007454

This book is produced from independently certified FSC™ paper to ensure responsible forest management.

For more information visit:
www.harpercollins.co.uk/green